KT-564-800

KEY CONCEPTS IN

HISTORICAL GEOGRAPHY

JOHN MORRISSEY
DAVID NALLY
ULF STROHMAYER
YVONNE WHELAN

s Angeles | London | New Delhi
gapore | Washington DC

Los Angeles | London | New Delhi
Singapore | Washington DC

SAGE Publications Ltd
1 Oliver's Yard
55 City Road
London EC1Y 1SP

SAGE Publications Inc.
2455 Teller Road
Thousand Oaks, California 91320

SAGE Publications India Pvt Ltd
B 1/I 1 Mohan Cooperative Industrial Area
Mathura Road
New Delhi 110 044

SAGE Publications Asia-Pacific Pte Ltd
3 Church Street
#10-04 Samsung Hub
Singapore 049483

© John Morrissey, David Nally, Ulf Strohmayer and Yvonne Whelan 2014

First published 2014

Apart from any fair dealing for the purposes of research or private study, or criticism or review, as permitted under the Copyright, Designs and Patents Act, 1988, this publication may be reproduced, stored or transmitted in any form, or by any means, only with the prior permission in writing of the publishers, or in the case of reprographic reproduction, in accordance with the terms of licences issued by the Copyright Licensing Agency. Enquiries concerning reproduction outside those terms should be sent to the publishers.

Editor: Robert Rojek
Editorial assistant: Keri Dickens
Production editor: Katherine Haw
Copyeditor: Neil Dowden
Marketing manager: Michael Ainsley
Cover design: Wendy Scott
Typeset by: C&M Digitals (P) Ltd, Chennai, India
Printed in Great Britain by Henry Ling Limited at The Dorset Press, Dorchester, DT1 1HD

Library of Congress Control Number: 2013944222

British Library Cataloguing in Publication data

A catalogue record for this book is available from the British Library

ISBN 978-1-4129-3043-7
ISBN 978-1-4129-3044-4 (pbk)

Neil Smith (1954–2012), *in memoriam*

CONTENTS

ABOUT THE AUTHORS

John Morrissey is a graduate of Trinity College Dublin, National University of Ireland and the University of Exeter. His research is primarily focused on questions of imperialism, geopolitics and resistance. He is the author of *Negotiating Colonialism* (2003) and co-editor of *Spatial Justice and the Irish Crisis* (2014). He previously taught at the University of Exeter and City University of New York, and now lectures at National University of Ireland, Galway. In 2012, he won the Irish National Academy Award for the Integration of Research, Teaching and Learning, and currently holds visiting research fellowships at Fitzwilliam College and Emmanuel College, Cambridge, where he is writing a geopolitical history of United States Central Command.

David Nally studied History and Geography at University College Cork before moving to Vancouver to pursue his doctoral degree at the University of British Columbia. He grew up in County Clare in the west of Ireland and is currently a Senior Lecturer in Human Geography and a Fellow of Fitzwilliam College at the University of Cambridge. His research interests include the political economy of agrarian change, the history of subsistence crises, and the politics of disaster relief. He has recently completed a book *Human Encumbrances: Political Violence and the Great Irish Famine* (2011) and is currently working on the making of the idea of 'food security'.

Ulf Strohmayer is a graduate of Munich Technical University and Pennsylvania State University. Currently, he is Professor of Geography at the National University of Ireland, Galway, after teaching previously at the University of Wales at Lampeter. Educated in Germany, Sweden, USA and France, he has also held visiting teaching and research posts at the Université de Pau et des Pays de L'Adour, Dresden Technical University, Binghamton University and the Maison des Sciences de l'Homme in Paris. His interest in social theory and philosophy is matched by an equal curiosity about the conditions and consequences of historical processes of modernization in Western Europe, all of which have informed his extensive publication record. He has also edited numerous volumes on social theory and the history of geographic thought.

Yvonne Whelan is a Senior Lecturer in Human Geography at the University of Bristol. She has published widely on the cultural landscapes of Ireland and is the author of *Reinventing Modern Dublin* (2003) and the co-editor of *Ireland: Space, Text, Time* (2005), *Heritage, Memory and the Politics of Identity* (2006) and *Ireland Beyond Boundaries* (2007).

ACKNOWLEDGEMENTS

Historical geography is fortunate to have such a rich and collegial academy, and it is a pleasure to recognize a wide range of wonderful colleagues here for their support, reading of chapters and patience as this book came to fruition. The patience required was due to the fact that the book has been a long time coming, for various reasons, but finally the acknowledgements get to be written, and written with pleasure.

John Morrissey would like to thank a number of colleagues who supported the project from the beginning. Particular thanks are due to Dan Clayton, Deborah Cowen, Zeynep Gambetti, Derek Gregory, Gerry Kearns, Stephen Legg, Alan Lester and Marilyn Young for their close reading of various chapters, and for their friendship and broader intellectual support over many years. Thanks are extended too to colleagues at NUI Galway for offering a wonderfully vibrant and supportive place to think, teach and write. Much of John's chapters were written while a research fellow at CUNY Grad Center, and his grateful thanks to colleagues there go to Jeff Bussolini, Gregory Donovan, Tina Harris, David Harvey, Cindi Katz and Ros Petchesky.

He hopes that former colleagues at CUNY do not mind that he singles out the late Neil Smith for individual acknowledgment. Neil was one of the most brilliant and inspiring geographers of his generation – indeed of any generation – and it still seems impossible that he is gone. His incisive political writing and activism both within and outside the academy reflected an unwavering commitment to questions of social justice, but in the many eulogies marking his passing it was sometimes forgotten that he was also a brilliant and accomplished historical geographer – his *American Empire* being one of the truly great books of historical geography in recent years. Neil looked forward to being at the launch of this book, and no doubt would tell us all to get on with it – to get on with the most pressing project of any time: to shake up the world with ideas and action.

David Nally wishes to express his gratitude to colleagues in the Department of Geography and Fitzwilliam College at the University of Cambridge. Special mention must be made to Ash Amin, David Beckingham, Chay Brooks, Simon Dalby, Simon Reid-Henry and Stephen Taylor for reading and commenting on early drafts of his chapters. At a crucial stage Phil Howell, in particular, set aside precious sabbatical time to provide detailed feedback and many pointers. In addition, he would also like to acknowledge Jim and Nancy Duncan, Matt Farish, Derek Gregory, Gerry Kearns and Willie Smyth for many years of constructive debate on the future, present and past of historical geography.

Ulf Strohmayer would like to thank his colleagues at NUI Galway for their ongoing support and shared sense of collegial workloads, which is key to combining individual research endeavours with equally important concerns of pedagogy and a functioning work environment. In addition, he extends his professional gratitude to colleagues abroad who have helped in the pursuit of empirical research that shaped his contributions to the present volume. These include, but are not limited to: the late Georges Benko, Dorothee Brantz, Paul Claval, Celina Cress, Tim Cresswell, Matt Hannah, David Harvey, Ilse Helbrecht, Judith Lazar, Julia Lossau, Gunnar Olsson, Aino Simon, the late Neil Smith, Ola Söderström, and Benno Werlen.

Yvonne Whelan would like to thank colleagues at the Department of Geography in University College Dublin who first introduced her to many of the concepts contained within the pages of this book and who inspired her interest in historical geography. In particular, she acknowledges Anngret Simms, Willie Nolan and Joe Brady. She would also like to thank the students she has taught at UCD and the Universities of Ulster, Montreal, Toronto and Bristol who helped to fine-tune her thoughts about historical geography.

Collectively, John, David, Ulf and Yvonne are also grateful for the support of the following friends and colleagues over the years: David Atkinson, Alison Blunt, Mark Boyle, Kate Brace, Mat Coleman, Nessa Cronin, Felix Driver, Paddy Duffy, Dave Featherstone, Diarmid Finnegan, Jennifer Fluri, Brian Graham, Dave Harvey, Mark Hennessy, Mike Heffernan, Nik Heynen, Alex Jeffrey, Rhys Jones, Innes Keighren, Rob Kitchin, James Kneale, David Lambert, Sharon Leahy, Mike Leyshon, Denis Linehan, Hayden Lorimer, Gordon MacLeod, Annaleigh Margey, Emma Mawdsley, Cheryl McEwan, Catherine Nash, Simon Naylor, Pat Nugent, Miles Ogborn, Sarah Radcliffe, James Ryan, Richard Smith, Tom Slater, Nicola Thomas, Karen Till, Gerard Toal, Bhaskar Vira and Charlie Withers. While the authors have drawn immense support from those mentioned they would like to be clear that they alone are responsible for any errors or faults that remain.

In addition, extended thanks are owed to Robert Rojek, Katherine Haw and Keri Dickens at SAGE. Their guidance and patience saw this project through in the most professional and supportive manner possible. And finally a sincere thank you to our dearest loved ones for reminding us every day about the things that really matter – for John: Olive and Darragh; for David: Estelle, Fergus and Eliza-Maeve; for Ulf: Christiane, Sebastian and Benjamin; and for Yvonne: Liam and Oisín.

John Morrissey
David Nally
Ulf Strohmayer
Yvonne Whelan

INTRODUCTION

HISTORICAL GEOGRAPHIES IN THE PRESENT

Introduction

'Those who cannot remember the past are condemned to repeat it' wrote Spanish-American writer, George Santayana (1980 [1905]: 284), just over a century ago. A similar theme was taken up by Karl Marx (1979 [1852]: 103) in *The Eighteenth Brumaire of Louis Napoleon*, where he too reasons that history repeats itself but adds the important sequitur 'first as tragedy, then as farce' (cf. Žižek, 2009). In both Marx and Santayana we are reminded that to ignore or forget the past is to risk becoming a prisoner of it. The Irish novelist James Joyce also explored this theme, but gave it a slightly different spin. In his celebrated novel *Ulysses*, Joyce gave one of his central characters Stephen Dedalus – a character loosely based on Joyce himself – the memorable line that 'history is a nightmare from which I am trying to awake' (1992 [1922]: 42). Writing at a time when his contemporaries were busy capturing the bewildering flux of modernity, Joyce depicted Dublin above all as a space of paralysis – a place where the past weighs heavily on the present ('[e]very Friday buries a Thursday', as Joyce creatively phrased it) and social relations are shaped, if never fully determined, by tidings from before (1992 [1922]: 138). Yet, there is another important but less explored side to Joyce's Dublin (cf. Kearns, 2006) that may be important in our present context. Perhaps more than any other modernist writer, Joyce celebrated the transformative effects of ordinary occurrences. Throughout *Ulysses*, hierarchy and authority are interrupted and undermined by the plurality, indeterminacy and inchoate anarchy of everyday practices. Of course there is both 'tragedy' and 'farce' in Joyce – and plenty of it – but there is also, we think, an unmistakable recognition of the *particular*, a celebration even of everyday events, and with it a profound appreciation that the thoughts and actions of ordinary people are never reducible to the frames of reference handed down to them from the past. Against the recurrent 'nightmare of history' Joyce positions the daydreams of ordinary people – their fears, desires, yearnings and hopes.

In this book we try to acknowledge the multiplicities of the past – but not in a manner that seeks to negate the deep asymmetries of both human history and historical representation. Rather these are core concerns we return to throughout. In the book we observe, on the one hand, how certain ideas, practices, modes of rule and configurations of authority are transposed in the present (Wacquant, 2009: 20). On the other hand, we are also acutely conscious of the need to avoid what E.P. Thompson (1991 [1963]: 12) termed the 'enormous condescension of posterity', which detects nothing meaningful in the actions of marginal groups who, having spent their lives on the fringes of mainstream society, are too often ignored in historical reflections. 'Only the successful (in the sense of those whose aspirations anticipated subsequent evolution) are remembered', continued Thompson (1991 [1963]: 12), '[t]he blind alleys, the lost causes, and the losers themselves are forgotten.' Approaching the past from the 'bottom up' is an essential step towards restoring the ideas, perspectives and aspirations of those that fall on the 'wrong side of history'. Their thoughts and deeds far from being obsolete remind us of the unassailable contingency of the present. 'Only that historian', wrote Walter Benjamin in his last major work *Theses in the Philosophy of History*, 'will have the gift of fanning the spark of hope in the past' (1992: 247; cf. Kaye, 2000). At a time when powerful interest groups and ideologues are claiming the 'end of history' and constructing dangerously reductive abstractions of geography (Kaplan, 2012), this sentiment has perhaps never been more important.

Thus an overarching methodological concern of this book is to ask geographical questions of the historical evidence that seeks to situate meaning in context. 'Historically sensitive geographies', remarks Matt Sparke (2007: 346–347), can teach us much about our contemporary world; today's 'authoritarianism in the Middle East', for example, emerges as a legacy of 'global historical geographies of uneven connection, exploitation, and oppression'. For Sparke, insisting on such historical geographies enables us to both denaturalize 'geographies of dispossession' and see 'authoritarianism' as a product of Western geopolitics rather than solely the result of 'native' character or 'tribal' geographies. Throughout this book, we endeavour to not only pay attention to 'historically sensitive geographies', but equally to 'geographically contextualized histories'. However, crucially, we wish to do so in a manner that reflects variously on our own positionalities, our own politics of historical representation. After all, as Felix Driver perceptively notes, historical scholarship cannot be detached 'from more worldly concerns with politics and ethics' (Driver, 2006: 1).

In researching and writing the worlds of the past, a key tradition that historical geography has developed over time is the ability to situate localized research in broader, comparative contexts. According to Brian Graham and Catherine Nash (2000: 1), historical geography possesses a particular strength in its dual concern for paying heed to 'both the specificity of the local and the wider economic, cultural and political processes and institutional structures'. In addition, historical geography has long comprised a variety of (often implied) theoretical and thematic concerns (Baker, 1987; Heffernan, 1997). More recently, new areas of research and a range of innovative methodological approaches have resulted in an even more vibrant 'pluralistic culture of scholarship' emerging, which not only builds upon existing sub-disciplinary strengths but has also become a 'driving force in the development of many of the new agendas of contemporary geography' (I.B.Tauris Publishers, 2009; see also Gagen et al., 2007). In this eclectic field of inquiry, it is vital of course that critical scholarship in historical geography does not become either diluted of politics or confined to merely textual analysis, but rather involves a *politics of scholarship* that engages broader concerns relating to the historical production of space and geographical knowledge (Lefebvre, 1991; Smith, 2003; Gregory, 2005).

Our open reference to and insistence on politics as an irreducible part of the construction and uses of geographical knowledge mirrors a wider acknowledgement in the humanities over the last 30 years or more of the constructed nature of disciplinary knowledges (see Heffernan, 2013). From its inception, historical geography as a sub-discipline was at the forefront of broader empiricist practices that sought to justify their existence and continued proliferation through a conscious engagement with *material* evidence of various kinds. Throughout the volume, we variously acknowledge the rich and manifold contributions made by this tradition. At the same time, however, we consciously situate our own modest endeavours in an additional set of theoretical and methodological practices aimed at locating historical geography within wider currents across the humanities and social sciences. Thus, we do not wish to use material artefacts, documents or other forms of historical evidence in a manner that reductively views time as the 'active' and space as the 'passive' element in historical explanations – a point brilliantly underscored by Michel Foucault (2007) in his well-known interview with the editors of the French journal *Hérodote*. Rather, we endeavour to account for past spatial configurations, processes and practices in a manner that contextualizes the *production* of knowledge as much as the knowledge itself.

If a reliance on positivist empiricism was traditionally implicated in claims to knowledge in the sub-discipline, we consider it to be imperative today that historical geographers engage self-critically with their various theoretical underpinnings and develop them with rigour, cohesion and discursive transparency. It is in this manner that we hope any insights here can begin to resonate beyond the walls of the sub-discipline and inform knowledges across the humanities respecting a broad range of normative concerns of, for instance, memory (Alderman and Inwood, 2013), resistance (Featherstone, 2008) and justice (Schein, 2011) – all of which can be fruitfully illuminated via critical historical geographical lenses.

Substantively, this book presents an overview of some of the most relevant and important concepts, practices and genealogies in the rich and vibrant field of historical geography. Our central aim has been to illuminate the relevancy of critical historical geography and of thinking in and across multiple temporal and spatial contexts. The final product of all historical geography is to (re)present some knowledge of the world from the past, to tell its story. This is not of course a straightforward exercise, as underlined throughout, but critically rendering visible the geographies of the past, accounting for their complexities and seeing their legacies can be a wonderfully rewarding and vitally important academic journey.

Section 1: Colonial and Postcolonial Geographies

The 'Colonial and Postcolonial Geographies' section explores one of the most significant, extensive and lasting historical geographical phenomena of modern times: colonialism. The initial chapter sets out the ideologies and imaginings of colonialism by exploring the key concepts of *imperialism* and *colonial discourse*. Particular focus is placed on imperialism as a concept and system of power, political economic ascendancy and cultural subordination, which relied on a networking of military, legal, and geopolitical power. Attention is then turned to interrogating how imperialism was imagined and scripted through colonial discourse by examining the dominant discursive registers that sought to normalize imperial mindsets and sustain the legitimacy of colonial expansion. The chapter outlines how historical geographers, influenced by the important work of Edward Said, have been especially proficient in alerting us to the subtle mechanisms of differentiation and purposeful relations of power, race, gender and sexuality inherent in the colonial discourses of former imperial powers. Finally, the chapter points to the historical role of geography, geographical techniques and geographers themselves in the advancement of imperialism. The subsequent chapter takes as its starting point the recognition of the importance of *discourse* in the justification of colonialism but then moves to interrogating how discourse became operationalized through colonial *practice* and governmentality. To this end, the key role that historical geography can play in studies of colonialism by locating analyses in necessarily grounded and differentiated ways is underlined. In examining practices of colonialism on the ground and highlighting the new geographies forged and contested as a result, focus is also placed on the political, economic and cultural practices of *anti-colonialism* that ensured the materialization of complex new spaces emerging under the shadow of colonialism. Historical accounts of the practices and spaces of colonial violence and anti-colonial resistance are shown to illuminate the echoes of the colonial past in the *colonial present*. The section is concluded with a chapter on *development* that seeks to demonstrate the imperial and colonial legacies of contemporary Western interventionism sanctioned under the aegis of development. It shows how at its core the concept of development internalizes what Wolfgang Sachs (2010: x) calls the need for 'backward peoples' to catch up with the 'pacemakers [of history] who are supposed to represent the forefront of social evolution'. As European states raced to acquire colonial possessions these ideals were channelled overseas as the 'will to improve' (Li, 2007) became an alibi for a 'humanitarian' mode of empire. It concludes by indicating how contemporary development, though now rooted in the language of technical and scientific innovation, still tends to formulate 'social evolution' in quintessentially Western terms.

Section 2: Nation-building and Geopolitics

The 'Nation-building and Geopolitics' section combines a number of important critiques of the role of geographical imagination, territoriality and spatiality in the historical construction of national identities and the expansionism of nation-states through geopolitical calculation. In the opening chapter 'Territory and Place', the notion of *territoriality* is explored; a term that historical geographers have used to describe the strategies used by people, groups or organizations to exercise power and control over a particular place and its component parts. The chapter concludes with a case study of the hotly contested and deeply segregated nature of territory in the north of Ireland, and examines some of the symbolic expressions of territoriality that prevail in such crucibles of conflict. In the subsequent chapter 'Identity and the Nation', geography is shown to have played a central role in the historical *construction, performance and reproduction* of national identity throughout the world. Using diverse examples, attention is initially focused on the importance of any nation's *public space* as a vital canvas through which to narrate and perform prioritized and selective meta-narratives of national identity. The chapter's key concern then shifts to problematizing reductive nationalist historical representations that mask key nuances from the geographical worlds of the past. In stressing how essentialist models of national identity promulgated *Otherness*, attention is especially drawn to the spatial mechanisms that facilitated exclusionary social practices and political and cultural hegemony. Finally, in underlining the historical social construction of all national identities, emphasis is placed on the selectivity of prioritized meta-narratives, and the importance of recognizing the historical relativism of all forms of absolute senses of identity whose endgame frequently involves racism, discrimination and conflict. The final chapter, 'Imaginative Geographies and Geopolitics', builds on this last point by examining the historical context of the use of *imaginative geographies* in the envisioning and representation of geopolitics, enmity and war. It uses the example of the current war on terror to reveal how *geopolitical calculation* involves the mobilization of *affective* imaginative geographies that serve to bury historical injustice, prior Western interventions and contemporary *geoeconomic interests* under a prevailing discourse of terror and threat. The argument is then made that historical geographers can play a vital role in historicizing and geo-graphing abstracted spaces and events. Emphasis is placed especially on how critical historical work in geopolitics can effectively challenge the legitimization and operation of contemporary Western geopolitical power/knowledge, and offer more nuanced *counter-geographies* that insist on the spatiality and materiality of historical and contemporary human geography.

Section 3: Historical Hierarchies

Taken together the chapters in this section advance three core claims about the historical geographies of identity making. First, it is argued that race, class and gender are best seen as concepts that take their meaning from their conjunction with other tropes and motifs and from the specific historical conditions under which they are deployed. In other words, it is crucial to appreciate the historical emergence of different 'epistemologies of difference' and to explain how they very often feed off and reinforce one another. The idea of race, for example, takes on different meanings and allusions depending on where, when and how it is deployed; like class and gender it is a concept that can be said to develop temporally and spatially. Second, the chapters show how these three 'epistemologies of difference' have influenced the field of human geography and more specifically the sub-field of historical geography. For instance, the traditional idea of geography as a male enterprise – and the prevailing notion of fieldwork as 'man's work' – has been roundly critiqued by feminist historical geographers. In challenging the myth of the active and autonomous male Self (the inimitable agent in the production of scientific knowledge who invariably prevails over the dependent and passive female Other imagined as the object of geographical inquiry) feminist scholars have shown how a theoretically informed study of geography's past(s) is a vital pre-condition for renewing the subject and for generating more inclusive geographies. The study of race, class and gender has contributed enormously to the study of how space is made, but equally and importantly it has profoundly shaped how geographers have themselves understood this enterprise. Finally, the chapters in this section actively explore how contemporary framings of difference rest upon, and are validated by, historical patterns of thought. Here it is claimed that historical hierarchies not only intersect and overlap, but also continue to inflect contemporary politics in very significant ways. Drawing attention to 'the historical geographies of the present' (Johnson, 2000), these chapters, like the volume itself, represent an attempt to show how supposedly 'natural' or 'common sense' claims are in fact historically produced world-views to which there have always been alternatives. In this sense forgotten, repressed and subaltern histories can provide a resource to think the present anew.

Section 4: The Built Environment

The relationship between people and space is the subject of scrutiny in Section 4. The opening chapter 'Nature and the Environment' examines the ways in which nature has become a topic of inquiry in historical geography, and explicitly addresses the shift that has taken place towards the natural world. In particular, this chapter conceptualizes *nature* as a product which is enmeshed in a wide range of contexts, from the gendering of social relations to the construction of both colonial and postcolonial identities. The remaining two chapters shift the emphasis towards the urban domain and focus attention on historical geographers' attempts to explore the complexity of urban spaces. These chapters place particular emphasis on the attempts that have been made to model *urban space* and document *urban morphology*. The chapter entitled 'Making Sense of Urban Settlement' reflects further on the processes and patterns that characterize urbanization, as well as on some of the trajectories that research has taken when it comes to making sense of one of the chief by-products of urbanization, namely the city and the townscape. The different ways in which geographers, sociologists and scholars from allied disciplines have attempted to make sense of cities are examined here, with particular emphasis on the seminal models of urban land use. The final chapter in this section considers one particular approach to the study of the urban landscape that was introduced to English-speaking urban historical geography by M.R.G. Conzen in the 1960s. 'Geographies of Urban Morphology' reviews Conzen's morphological method and privileges a more specific concern with the built fabric and the urban morphology or form of the city. The chapter foregrounds the important concepts and methodological techniques developed by Conzen and which continue to aid historical geographers in their work on the geographical character of towns and cities.

Section 5: Place and Meaning

In Section 5 of the book, a wide range of issues relating to place and meaning come into view. In particular, the concepts treated here address some of the ways in which geographers operating at the boundaries between cultural and historical geography have conceptualized landscape, memory and heritage. Using a range of case studies, this section begins with an account of *landscape* and outlines some of the ways in which it has been studied over time. As well as reflecting on the development of the landscape concept in human geography, this chapter considers the approaches to landscape study that developed out of the 'new cultural geography' in the 1990s. In particular, it examines the symbolic geography of the cultural landscape and the rich nexus of inquiry that has coalesced around landscape, memory and identity. The political, economic and cultural uses of the past are also examined in the chapter on *heritage* that follows. This chapter asks what do we really mean when we talk about heritage, and goes on to explore some of the ways in which it has been conceptualized. As well as reviewing the political, economic and social dimensions of the heritage discourse, the chapter pays particular attention to a case study of Irish famine heritage in order to probe further the pivotal role of heritage in shaping narratives of diasporic identity. This section concludes with the chapter 'Performance, Spectacle and Power', which interrogates the historical geographies of spectacles, parades and public performances and reviews some of the ways in which historical geographers have studied the role of public parades as rituals of remembrance, as well as choreographed expressions of power. The chapter foregrounds the multi-faceted impact of public parades and *performance*, and traces the ways in which this impact is mediated materially and militarily through the skilful appropriation of aspects of the past, as well as through pageantry, illuminations, fanfare and music. This chapter concludes with a brief reading of one such public spectacle, the much contested visit of Queen Victoria to Ireland in 1900.

Section 6: Modernity and Modernization

A key concept in any analysis of the recent past, the notion of 'modernity' and its accompanying adjective have arguably long held pride of place among geographical research and scholarship devoted to historically resonating themes. Even though the term lacks both precision and a priori content, its contextual allure continues to entice and will be the centre of this section's attention. In its wake, the chapter will analyse the construction of historical geographies attaching to some of the material and ideal transformations customarily thought to fill 'modernity' with meaning. Of these, processes associated with and characterized by capitalism will be the first to be accorded scrutiny, allowing us to stress the extended historical trajectory and geographical specificities attaching to this seminal development in the historical geography of humankind. In a similar vein, the chapter will focus on the industrial revolution and urbanization as interrelated processes that warrant historico-geographical analyses. Subsequent chapters move the focus from productionist processes anchored in real economies to those associated with knowledge and science, acknowledging that these, too, contribute centrally to the ongoing history of modernity. Here we set out to question widely used notions such as the 'scientific revolution', 'the enlightenment' or the deployment of teleological modes of reasoning more generally within contemporary accounts of modernity. A central tenet in these accounts has become the work accomplished through 'networks' of various kinds, rendering the many possible connections between writing, the imagination and economic practices, as well as the sites and technologies through which they are established, objects of geographical curiosity. The final chapter of the section turns its attention to the accompanying normative project attached to notions of modernity, summarised succinctly underneath the rubrics of 'democracy' and of the 'public sphere'. Arguing for a broadening of the concept of modernity in terms of practical work and research being carried out, this chapter seeks to establish more cogent linkages between space, place, sites and the construction of specifically 'modern' forms of legitimacy.

Section 7: Beyond the Border

The penultimate section of the book addresses 'big picture' themes in historical geographical research. The first chapter on globalization begins with the vexed question of periodization: when does globalization begin and how do we define its stages of evolution? These are questions that have exercised historical geographers (and other social scientists) for a long time and while there is no established consensus, we argue along with Arjun Appadurai (1996) that contemporary globalization has significant 'precursors' and 'sources' in the past and it is only through a deeper understanding of such historical precedents that we can begin to come to terms with the nature and dynamic of these global-scale changes in the present. If globalization marks a longstanding preoccupation of historical geographers, the theme of the next chapter in this section, 'Governmentality', addresses a more recent concern of historical geographers. Drawing on a work of French thinker Michel Foucault, historical geographers have turned their attention to analysing what has been described as the 'spatial ontology' of power (Philo, 2000: 218). Foucault is unusual among European philosophers for insisting that space is not simply the stage on which social relations 'play out', but is actually constitutive of those self-same relations. This chapter shows how historical geographers have developed this core idea through a range of empirical case studies. The final chapter focuses on 'social nature' and considers how the material world has been modulated and changed over time. This concern runs to the core of the discipline of geography, but it is of central importance to historical geographers who add to the 'socio-cultural' interpretation of landscapes, and the 'political' reading of ecology, an historical framing of 'nature' that is at once alive and sensitive to the transformations pressed on the world by human and non-human actors.

Section 8: The Production of Historical Geographical Knowledge

The final section of the book explores the conditions of possibility for historical geographical scholarship to emerge. Given the title and remit of the book, the first target of such an undertaking must surely be the very notion (as well as the associated practices) of historical geography itself. As sub-disciplines go, this one is a fairly eclectic and broad one; even so, it defines what it is through practices of inclusion and exclusion, inevitably naturalizing the pursuit of certain forms of knowledge. In this section, we focus on the epistemological coining of concepts informing the production of knowledge in historical geography, while also contextualizing and deconstructing some of its key techniques, discourses and manners of thinking more generally. Not surprisingly, the notion of 'evidence' is central to this endeavour – and it bears analysing in detail how evidence is variously constructed and theoretically informed. The section thus considers questions of fieldwork, sources and cartography in the context of a broader set of ordering practices within the discipline of geography. The particular emphasis on visual methodological practices is not accidental given their centrality in the annals of historical geography. The sub-discipline has long used and critiqued 'the visual' as a means of communication. Finally, the section variously considers the broader import of historical geographical critique. Given both its temporal and spatial concerns, historical geography is perhaps uniquely poised to expound critical contextualized analyses of the worlds of the past and, in doing so, to offer important readings too of the legacies of those worlds in the present.

References

Alderman, D.H. and Inwood, J. (2013) 'Street naming and the politics of belonging: spatial injustices in the toponymic commemoration of Martin Luther King Jr', *Social & Cultural Geography*, 14(2): 211–233.

Appadurai, A. (1996) *Modernity at Large: Cultural Dimensions of Globalization*. Minneapolis: University of Minnesota Press.

Baker, A.R.H. (1987) 'Editorial: the practices of historical geography', *Journal of Historical Geography*, 13(1): 1–2.

Benjamin, W. (1992) *Illuminations* (edited with an introduction by Hannah Arendt; translated by Harry Zohn). London: Fontana Press.

Driver, F. (2006) 'Editorial: historical geography and the humanities: more than a footnote', *Journal of Historical Geography*, 32(1): 1–2.

Featherstone, D. (2008) *Resistance, Space and Political Identities: The Making of Counter-Global Networks*. Oxford: Blackwell.

Foucault, M. (2007) 'Questions on geography', in J. Crampton and S. Elden (eds) *Space, Knowledge and Power*. Farnham: Ashgate, pp. 173–184.

Gagen, E., Lorimer, H. and Vasudevan, A. (eds) (2007) *Practising the Archive: Reflections on Method and Practice in Historical Geography*. London: HGRG, Royal Geographical Society.

Gregory, D. (2005) 'Geographies, publics and politics', *Progress in Human Geography*, 29(2): 182–193.

Heffernan, M. (1997) 'Editorial: the future of historical geography', *Journal of Historical Geography*, 23(1): 1–2.

Heffernan, M. (2013) 'Geography and the Paris Academy of Sciences: politics and patronage in early 18th-century France', *Transactions of the Institute of British Geographers* (early view, doi: 10.1111/tran.12008).

I.B.Tauris Publishers (2009) 'Tauris historical geography series', www.ibtauris.com/pdf/historicalgeographyflyer.pdf (accessed 14 July 2009).

Johnson, N. (2000) 'Historical geographies of the present', in B. Graham and C. Nash (eds) *Modern Historical Geographies*. Harlow: Prentice Hall, pp. 251–272.

Joyce, J. (1992 [1922]) *Ulysses*. London: Penguin.

Kaplan, R. (2012) *The Revenge of Geography: What the Map Tells Us About Coming Conflicts and the Battle Against Fate*. New York: Random House.
Kaye, H (2000) 'Fanning the spark of hope in the past: the British Marxist historians', *Rethinking History: The Journal of Theory and Practice*, 4(3): 281–294.
Kearns, G. (2006) 'The spatial poetics of James Joyce', *New Formations*, 57: 107–125.
Lefebvre, H. (1991) *The Production of Space* (trans. D. Nicholson-Smith). Oxford: Blackwell.
Li, T.M. (2007) *The Will to Improve: Governmentality, Development, and the Practice of Politics*. Durham, NC: Duke University Press.
Marx, K. (1979) [1852] *The Eighteenth Brumaire of Louis Napoleon*, in K. Marx and F. Engels, *Collected Works, XI*. London: Lawrence & Wishart.
Nash, C. and Graham, B. (2000) 'The making of modern historical geographies', in B. Graham and C. Nash (eds) *Modern Historical Geographies*. Harlow: Prentice Hall, pp. 1–9.
Philo, C. (2000) 'Foucault's Geography' in M. Crang and N. Thrift (eds) *Thinking Space*. London: Routledge, pp. 205–238.
Sachs, W. (ed.) (2010) *The Development Dictionary: A Guide to Knowledge as Power*. 2nd edition. London: Zed Books.
Santayana, G. (1980) [1905] *The Life of Reason, Volume 1* (first published Charles Scribner's Sons, New York). New York: Dover Publications.
Schein, R. (2011) 'Life, liberty and the pursuit of historical geography', *Historical Geography*, 39: 7–28.
Smith, N. (2003) *American Empire: Roosevelt's Geographer and the Prelude to Globalization*. Berkeley: University of California Press.
Sparke, M. (2007) 'Geopolitical fears, geoeconomic hopes, and the responsibilities of geography', *Annals of the Association of American Geographers*, 97(2): 338–349.
Thompson, E.P. (1991) [1963] *The Making of the English Working Class*. London: Penguin.
Wacquant, L. (2009) *Punishing the Poor: The Neoliberal Government of Social Insecurity*. Durham, NC: Duke University Press.
Žižek, S. (2009) *First as Tragedy, Then as Farce*. London: Verso.

Section 1 Colonial and Postcolonial Geographies

1 IMPERIALISM AND EMPIRE

John Morrissey

Introduction

From the expeditions to the Americas in the fifteenth century through to the interventions in the Middle East in the twenty-first century, imperialism has indelibly marked modern times, forging human geographies on every continent and leaving legacies still seen and lived to this day. As Robert Young notes, the 'entire world now operates within the economic system primarily developed and controlled by the West, and it is the continued dominance of the West, in terms of political, economic, military and cultural power that gives this history a continuing significance' (2001: 5). Moreover, given the ongoing wars prosecuted in the name of Western civilization in the world today, Derek Gregory prompts us to recognize 'the ways in which so many of us continue to think and act in ways that are dyed in the colors of colonial power' (2004: xv). This chapter initially sets out the ideologies and discursive mobilities of imperialism before reflecting on approaches to its study in historical geography, while the subsequent chapter examines the complex geographies of colonialism and anti-colonialism forged and contested throughout the world as a result.

Defining Imperialism

Imperialism can be defined as a system of power, political economic ascendancy and cultural subordination, envisioned from the centre of expanding nation-states and differentially operationalized in colonized spaces throughout the world. Definitions are always fraught with difficulties, of course, and it is important to recognize the complexities of the terms *empire*, *imperial* and *imperialism*, which have been shown to have connoted different historical cultural meanings and political realities through time (Loomba, 1998). In general terms, imperialism has historically operated in various forms. As Dan Clayton outlines, there have been 'over 70 empires in history' (2009a: 189). Temporally, these comprise ancient, medieval, early modern, modern and contemporary, and geographically include, for example, the former Inca, Greek, Roman, Chinese, Ottoman, Spanish, British, Japanese and Soviet empires. Differentiating models of empire is hugely problematic; there has been considerable debate concerning the extent to which colonial expansion was state-driven and centred, for example (Hardt and Negri, 2000). That said, three key variants of *state-driven* imperialism on a *global* scale are: (a) the early modern Spanish imperial model; (b) the more globalized and advanced version of the major European powers of the late nineteenth century; and (c) the new imperialism or neo-imperialism of US military and economic ascendancy in the present (Johnson, 2000; Young, 2001; Harvey, 2003; Smith, 2003; Gregory, 2004).

Colonizations took place in Europe, Asia and elsewhere in the medieval and earlier periods, when the Greek, Roman, Chinese and Islamic empires advanced in geographically contiguous territories but largely without specific mercantile or state-driven logics of expansion. The first modern, transoceanic and state-driven global empire, however, was forged in the New World of the Americas by the conquering armies of the Spanish conquistadors from the late fifteenth century. The bureaucratic Spanish administrations in these new worlds were typically dependent on isolated military power and direct taxation on indigenous peoples and were not initially at least integrated into an imperial network of capitalist overseas endeavours like later European empires (Young, 2001).

Imperialism in its nineteenth-century design was developed by the French via the notion of a *mission civilisatrice*, which was an ideological justification for aggressive territorial expansion enabled by technological innovation. The *mission civilisatrice* invoked the idea of bringing French civilization, culture and language, together with Christianity, to the uncivilized and unenlightened, who were to be assimilated. This neat justification for superimposing the cultures and values of *us* on *them* was also a key feature in the contemporary British notion of a *civilizing mission*. However, both ideologies of empire had previous antecedents in early modern Spanish and English colonial discourses in the sixteenth and seventeenth centuries that centred on notions of reform and assimilation.

By the late nineteenth century, the French, British and other European imperial powers were 'increasingly drawn into a competitive global economic and political system', whose central underlying objective was to 'combine the provision of domestic political and economic stability with the production of national prestige and closed markets in the international arena through conquest' (Young, 2001: 30–31). According to Young, the imperial scramble for Africa in the early twentieth century by the British, French, Germans and others represented the high point of imperial state rivalries and reflected an expanded capitalist world economy, typified by increased production and consumption. Young's tendency to see imperial growth as almost exclusively state-driven, however, ignores the multiplicity of interests and projects pursued by Europeans that might ultimately result in formal or indeed informal imperialism. For example, there was no state logic to the Puritan colonization of America, the missionary-led colonization of the Pacific, or (directly at least) the East India Company's activities in India (Lambert and Lester, 2004).

Imperialism in its formal sense effectively ended with the retreat of the European empires as the twentieth century progressed, and this was due to a number of factors including: the Bolshevik revolution in Russia and the emergence of a powerful state opposed to Western imperialism; resistance to empire from colonized peoples throughout the world; the growing inability of European powers to administer their colonies effectively after the exhaustions and expense of World War II; and finally the subsequent appearance of a new superpower on the world stage, the USA, which viewed existing imperial trading structures as an impediment to its own economic activities overseas (Young, 2001; Larsen, 2005). The last reason cited here points to the fact that the new world order that replaced imperialism was in many ways a more subtle, informal version of the same favourable economic power structures dictated by the West – often referred to as *neo-imperialism*.

Imperialism and Discourse

Imperialism was legitimized and sustained through purposeful discursive imaginings, identifications and ascriptions, referred to as *colonial discourse* (also typically referred to as *imperial discourse*, but for the purposes of clarity in this chapter and the next, the term *colonial discourse* is used). Its analysis is critical to our understanding of how imperialism works (see the next chapter for its overarching relations to colonial practice). Colonial discourse equates to the prevailing representations of imperial power that sought to normalize imperial mindsets and the legitimacy of colonial intervention and domination. Imperialism should not be understood as only driven by political and economic logics, as Nicholas Thomas reminds us; rather, it 'has always, equally importantly and deeply, been a cultural process' in which 'discoveries and trespasses are imagined and energized through signs, metaphors and narratives' (1994: 2). In other words, imperialism's cultural discourses served not simply to 'mask, mystify or rationalize forms of oppression that are external to them' but were 'constitutive of colonial relationships in themselves' (ibid.: 2). This is what Loomba means when she asserts that 'power works through language, literature, culture and the institutions which regulate our daily lives' (1998: 47).

In exploring questions of colonial discourse, a fundamental starting point is Edward Said's illuminating and seminal work *Orientalism* (1978). Inspired by Michel Foucault's work on the intrinsic relationship between power and knowledge, *Orientalism* examined a wide range of Western representations of the East by novelists, academics and others during the course of the nineteenth and twentieth centuries, which Said showed to create a collective, powerful European imaginary of the Orient in which the West is posited via a series of binaries as a superior, civilized and rational authority over an inferior, barbaric and irrational subordinate. For Said, imperialism was underpinned by these powerful discourses or representations of what he called *Otherness*. Said's notion of *imaginative geographies* (see also Chapter 6), which he revealed to be inherent in the colonial discourse of Orientalism, was further elaborated in his work *Culture and Imperialism* in 1993. As Karen Morin highlights, this key concept has revealed the 'invention and construction of geographical space' that 'constructs boundaries around our very consciousness and attitudes, often by inattention to or the obscuring of local realities', and to this end the concept has been hugely significant in drawing careful attention to 'spatial sensitivity' in colonial and postcolonial studies (2004: 239).

In thinking through how discursive modalities function in the identification of Otherness, Said's notion of *cultural ascription* is particularly useful. He has shown how the prioritization and networking of the language and representational practices of Orientalism, for example, serves to collectively naturalize human 'types' via 'scholarly idioms and methodologies', which literally 'ascribes reality and reference' (1978: 321). Colonial discourse generates and sustains dominant and colonizing knowledges

(knowledges that in turn facilitate power) by ascribing identity and difference to distinct spaces, places and peoples (Gregory, 2001; see also Chapter 5). Colonial discourse's binaries of Self and Other are ultimately cemented by institutional ascriptions of human *types* to specific environmental and cultural settings (see Chapter 23 for a discussion of map-making and geography's role as a discipline in colonial history).

Inspired and informed by Said's postcolonial critique, various geographers have alerted us to the subtle mechanisms of differentiation and purposeful relations of power, race, gender and sexuality inherent in the colonial discourses of former imperial powers (Blunt and Rose, 1994; Lester, 2001; Morrissey, 2003; Clayton, 2004; Kumar, 2006). The histories of geography have also been examined in recent years, highlighting the role of geographical institutions, methods and academics themselves in imperial practices of 'exploration, mapping and landscape representation, and divisive discourses on climate and race' (Clayton, 2009a: 190; see also Ploszajska, 2000 and Heffernan, 2003). Many geographers have sought particularly to 'decolonize the geographical constitution and articulation of colonial discourses in both the past and the present, [and] also to decolonize the production of geographical knowledge both in and beyond the academy' (Blunt and McEwan, 2002: 1). In addition, however, they have also 'warned against reducing imperialism to discourse' and have insisted on 'the need to materially ground understanding of imperialism's operations' (Clayton, 2009b: 374; this is the focus of the next chapter).

Approaches to Understanding Imperialism

One of the challenges in studying colonial geographies is that of drawing the sometimes problematic conceptual distinction between imperialism and colonialism. Robert Young's luminous work *Postcolonialism* is particularly instructive on this primary point. Young makes the useful argument that *imperialism* can be equated to a *concept* or ideology of territorial expansion, economic control and cultural superiority, while *colonialism* is best understood as the *practice* of domination of alien peoples, frequently though not always underpinned by imperialism. For Young, imperialism was 'typically driven by ideology from the metropolitan centre and concerned with the [systematic] assertion and expansion of state power', whereas colonialism was primarily 'economically driven' by migrant settler communities, speculators or trading companies and was concerned with more ad hoc, localized matters of territorial and economic administration (Young, 2001: 16–17). Putting the distinction between imperialism and colonialism another way, Ania Loomba prompts us to think of the difference between them in 'spatial terms', where imperialism 'originates in the metropolis' and leads to the process of 'domination and control', while its effect, colonialism, is what 'happens in the colonies as a consequence of imperial domination' (Loomba, 1998: 6). Of course, like all models, Loomba's albeit useful working distinction is complicated by the 'local'. The study and writing of historical geography requires a careful attentiveness to context, and it is not just the 'local' that requires theorising; paying attention to the 'transnational' elements of imperialism can also be crucial, and this adds a further conceptual challenge, as Stephen Legg (2010) reminds us.

In historical geography, the study of imperialism has been critically approached in at least three main ways, as Dan Clayton (2009b: 373–374) has shown: first, imperialism has 'been analysed in economic and political terms – as central to the evolution of capitalism and the nation-state' (Lenin theorized imperialism as the 'highest stage of capitalism' (Lenin, 1969)); second, since the 1980s, imperialism 'has been studied as a discourse – or grammar – of domination fuelled by images, narratives and representations, and shaped by categories of gender, sexuality, race, nation and religion, as well as capital and class'; and finally, imperialism has more recently been examined via an '"imperial networks" approach', which 'treats metropole and colony as mutually constitutive'. All three approaches have been critiqued in relation to how effectively they discriminate and disaggregate different logics of imperial power. Harris (2004: 165), for example, has charged approaches that concentrate on imperialism as a discourse with privileging the cultural logic of imperialism and therefore obscuring 'other forms of colonial power'.

For the contemporary world, neo-imperialism has been identified and studied in two further interrelated ways: (a) in the Marxist sense as a system of economic domination, frequently associated with the West's hegemonic world position; and (b) via the notion of an American military and economic *empire lite* or *empire in denial*; the advancement of which accelerated under the pretext of the so-called *war on terror* in the post-9/11 world (Agnew, 2003; Harvey, 2003; Ignatieff, 2003; Gregory, 2004; Larsen, 2005; Smith, 2005). In examining neo-imperialism, many historical geographers have offered insightful critiques by carefully spatializing and historicizing the antecedents of contemporary Western interventions overseas. A key strength of any critical historical geography lies in its capacity to contextualize the present by recognizing the past – seeing its legacies and narrating the historical relations of power and politics that continue to bound people and places throughout the world (see, for example, Kearns, 2006).

Conclusion

By way of conclusion, I want to return to the work of Edward Said. As the key writer in the field, his reading of imperialism and discourse is vitally important. It can be challenged by the fact that it concentrated almost solely on the written text of high culture to the detriment of the various visual cultural discourses of art and photography and other representations such as travel writing. Said's relative lack of attention to gender and his sometimes limited acknowledgment of the agency of anti-colonial resistance have also come in for specific criticism (Lester, 2000; Young, 2001). However, his critical deconstruction of the historical language, power relations and subject positions of the Western *Self* and the external *Other* continues to have an enduring legacy. His work still possesses a key relevance to the world today in which crude and essentialist distinctions between *us* and *them* are continuously invoked in the all-powerful and omnipresent discourse of the *war on terror*. Today, sadly as much as ever, stereotyped representations frequently stand for knowledge itself and underpin the execution of power and violence.

KEY POINTS

- Imperialism can be defined as a system of power, political economic ascendancy and cultural subordination, envisioned from the centre of expanding nation-states and differentially operationalized in colonized spaces throughout the world.
- State-driven imperialism on a global scale has historically operated in three principal forms: (a) the Spanish imperial model; (b) the more globalized and advanced version of the major European powers of the late nineteenth century; and (c) the new imperialism or neo-imperialism of US military and economic ascendancy in the present.
- In the mid-nineteenth century, the French notion of a *mission civilisatrice* and British notion of a *civilizing mission* were soon adapted by

the other European imperial powers, and the early twentieth-century scramble for Africa represented the high point of imperial rivalries in an expanded capitalist world economy.

- Imperialism was imagined, legitimized and sustained through *colonial discourse*, which equates to the prevailing representations of imperial power that sought to normalize imperial mindsets and the rights of colonial expansion and domination.
- Inspired and informed by Edward Said's postcolonial critique, historical geographers have alerted us to the subtle mechanisms of differentiation and purposeful relations of power, race, gender and sexuality inherent in the colonial discourses of former imperial powers.
- Critical histories of geography reveal the role of geographical institutions, methods and academics in the advancement of imperialism.

References

Agnew, J. (2003) 'American hegemony into American empire? Lessons from the invasion of Iraq', *Antipode*, 35(5): 871–885.

Blunt, A. and McEwan, C. (eds) (2002) *Postcolonial Geographies*. London: Continuum.

Blunt, A. and Rose, G. (eds) (1994) *Writing Women and Space: Colonial and Postcolonial Geographies*. New York: Guilford.

Clayton, D. (2004) 'Imperial geographies', in J.S. Duncan, N.C. Johnson and R.H. Schein (eds) *A Companion to Cultural Geography*. Oxford: Blackwell, pp. 449–468.

Clayton, D. (2009a) 'Empire', in D. Gregory, R. Johnston, G. Pratt, M. Watts and S. Whatmore (eds) *The Dictionary of Human Geography* (5th edn). Chichester: Wiley-Blackwell, pp. 189–190.

Clayton, D. (2009b) 'Imperialism', in D. Gregory, R. Johnston, G. Pratt, M. Watts and S. Whatmore (eds) *The Dictionary of Human Geography* (5th edn). Chichester: Wiley-Blackwell, pp. 373–374.

Gregory, D. (2001) '(Post)colonialism and the production of nature', in N. Castree and B. Braun (eds) *Social Nature*. Oxford: Blackwell, pp. 84–111.

Gregory, D. (2004) *The Colonial Present: Afghanistan, Palestine, Iraq*. Oxford: Blackwell.

Hardt, M. and Negri, A. (2000) *Empire*. Cambridge, MA: Harvard University Press.

Harris, C. (2004) 'How did colonialism dispossess? Comments from an edge of empire', *Annals of the Association of American Geographers*, 94(1): 165–182.

Harvey, D. (2003) *The New Imperialism*. Oxford: Oxford University Press.

Heffernan, M. (2003) 'Histories of geography', in S.L. Holloway, S.P. Price and G. Valentine (eds) *Key Concepts in Geography*. London: Sage, pp. 3–22.

Ignatieff, M. (2003) *Empire Lite: Nation-Building in Bosnia, Kosovo and Afghanistan*. Toronto: Penguin.

Johnson, C. (2000) *Blowback: The Costs and Consequences of American Empire*. New York: Metropolitan.

Kearns, G. (2006) 'Naturalising empire: echoes of Mackinder for the next American century?', *Geopolitics*, 11(1): 74–98.
Kumar, S. (2006) 'The census and women's work in Rangoon, 1872–1931', *Journal of Historical Geography*, 32(2): 377–397.
Lambert, D. and Lester, A. (2004) 'Geographies of colonial philanthropy', *Progress in Human Geography*, 28(3): 320–341.
Larsen, N. (2005 [2000]) 'Imperialism, colonialism, postcolonialism', in H. Schwarz and S. Ray (eds) *A Companion to Postcolonial Studies*. Oxford: Blackwell, pp. 23–52.
Legg, S. (2010) 'Transnationalism and the scalar politics of imperialism', *New Global Studies*, 4(1): 1–7.
Lenin, V.I. (1969 [1916]) *Imperialism: The Highest Stage of Capitalism*. New York: International Publishing Company.
Lester, A. (2000) 'Historical geographies of imperialism', in B. Graham and C. Nash (eds) *Modern Historical Geographies*. Harlow: Prentice Hall, pp. 100–120.
Lester, A. (2001) *Imperial Networks: Creating Identities in Nineteenth Century South Africa and Britain*. London: Routledge.
Loomba, A. (1998) *Colonialism/Postcolonialism*. London: Routledge.
Morin, K. (2004) 'Edward W. Said', in P. Hubbard, R. Kitchin and G. Valentine (eds) *Key Thinkers on Space and Place*. London: Sage, pp. 237–244.
Morrissey, J. (2003) *Negotiating Colonialism*. London: HGRG, Royal Geographical Society.
Ploszajska, T. (2000) 'Historiographies of geography and empire', in B. Graham and C. Nash (eds) *Modern Historical Geographies*. Harlow: Prentice Hall, pp. 121–145.
Said, E. (1978) *Orientalism: Western Conceptions of the Orient*. New York: Pantheon.
Said, E. (1993) *Culture and Imperialism*. New York: Alfred A. Knopf.
Smith, N. (2003) *American Empire: Roosevelt's Geographer and the Prelude to Globalization*. Berkeley: University of California Press.
Smith, N. (2005) *The Endgame of Globalization*. New York: Routledge.
Thomas, N. (1994) *Colonialism's Culture: Anthropology, Travel and Government*. Cambridge: Polity.
Young, R.J.C. (2001) *Postcolonialism: An Historical Introduction*. Oxford: Blackwell.

FURTHER READING

Clayton, D. (2004) 'Imperial geographies', in J.S. Duncan, N.C. Johnson and R.H. Schein (eds) *A Companion to Cultural Geography*. Oxford: Blackwell, pp. 449–468.
Lester, A. (2000) 'Historical geographies of imperialism', in B. Graham and C. Nash (eds) *Modern Historical Geographies*. Harlow: Prentice Hall, pp. 100–120.
Said, E. (1978) *Orientalism: Western Conceptions of the Orient*. New York: Pantheon.
Young, R.J.C. (2001) *Postcolonialism: An Historical Introduction*. Oxford: Blackwell.

2 COLONIALISM AND ANTI-COLONIALISM

John Morrissey

Introduction

The range of ideologies of even one former imperial power and the manner in which its colonial modalities were differentially applied in varying geographical and historical contexts makes it difficult to simplify any grand theory of colonialism. However, if imperialism was imagined and legitimized via different discursive logics or discourses, its manifestation in practice typically resulted in expropriation, violence and resistance to the imposition of political and cultural values over alien peoples. This chapter interrogates the historical geographies of colonialism by connecting colonial *discourse* to colonial *practice* and by examining the profound impact colonial interventions had, and continue to have, on people and places on every continent.

Thinking Through Colonialism: Discourse and Practice

Colonialism can be thought of as a distinct Western modality of power, an intrinsically exploitative and dehumanizing system of control. As Dan Clayton puts it, it can be viewed as 'symptomatic of an epistemological malaise at the heart of Western modernity – a propensity to monopolise and dictate understanding of what counts as right, normal and true, and denigrate and quash other ways of knowing and living' (2009: 94). As explored in the previous chapter, this system was envisioned through discourse. However, it is important to not only recognize the import of colonial *discourse* in the justification of imperialism but also to see its very real connections to colonial *practice* (see Table 2.1 for a broad sketch of some of the connections). By way of illustration, I want to use here the example of England's first early modern colony, Ireland. The beginning of the English colonial project in Ireland in the sixteenth century was preceded and underpinned by a long sequence of specific geographical discourses that effectively *Othered* the existing Gaelic-Irish population as barbaric, unlawful and in desperate need of civilizing. Such treatises were advanced by leading contemporary English government officials and travel writers in Ireland. Consider, for example, one such individual, John Derricke, whose pictorial and narrative depiction of Gaelic lawlessness, cattle-raiding and house-burning in 1581 (seen in Figure 2.1) formed part of a wider series of contemporary discursive representations that served to demonize the Gaelic-Irish and simultaneously garner the Elizabethan royal court's ideological and economic backing for the Munster Plantation in the south of the country. Some ten years later, the endgame was a new colonial economy brought about by colonial violence, dispossession and settlement, and buttressed by a new military and political order (Morrissey, 2003).

In the various civilizing missions of the main European imperial powers of the nineteenth century (sometimes referred to as the age of *high colonialism*), essentialist geographical discourses, underpinned by Western conceptions of order and truth, continued to feature centrally (Mitchell, 1991). Imperial interventions in the contemporary world still rely, of course, on reductive geographical formulations. In one of his last works before he died in 2003, Edward Said wrote of the import of Manichean geographical knowledges in the latest *mission civilisatrice* of the war in Iraq, appealing to his readers not to 'underestimate the kind of simplified view of the world that a relative handful of Pentagon civilian elites have formulated for US policy in the entire Arab and Islamic worlds' (2003: xx). For Said, the dreadful consequences of a long-established Western triumphalist discourse of *us* and *them* underpinned the outbreak of the war: 'Without a well-organized sense that these people over there were not like "us" and didn't appreciate "our" values – the very core of traditional Orientalist dogma [or Western colonial discourse] – there would have been no war (2003: xv)'.

Notwithstanding Edward Said's invigorating and irrevocable impact on studies of colonial geographies, his work has tended to homogenize and generalize the discourses and practices of Western imperial powers. As Alan Lester argues, 'the image of an overarching metropolitan representation of other places and peoples, or a uniform European agenda, needs

Table 2.1 Imperialism, discourse and colonialism

Imperialism

- Empire and interventionism
- Empire and economics
- Empire and nation-building

↕

Imperial/Colonial Discourse

- Power/knowledge couplet
- Representational and performative binaries
- Regimes of truth

↕

Colonialism

- Dispossession, settlement and capitalist accumulation
- Discrimination, biopolitics and regulation
- Contact zones, resistance and violence

Figure 2.1 The demonisation of the Gaelic-Irish

Source: John Derricke's *Image of Irelande, 1581* (reproduced courtesy of the National Library of Ireland)

to be disaggregated' (Lester, 2000: 102). Much work in geography has also tended to affirm primacy to the Western metropole or centre in exploring colonialism, which runs the risk of reifying notions of colonial Eurocentrism and overstating the spatial and ideological separations between core and periphery (for a largely metropolitan focus, see for example Godlewska and Smith, 1994). Using the example of the British Cape Colony in nineteenth-century southern Africa, Lester (2002) has carefully divulged on the contrary the discursive connections and co-constitution of emerging colonial discourses of race, class and cultural subordination that operated across a trans-imperial network between Britain – as the metropole and centre of imperial power – and the Cape – as a colony and site of colonial practice. In recent years, more sustained engagement with the particularity of the discourses and practices of colonialism in the colonized worlds themselves has emerged (see, for example, Morrissey, 2003; Raju et al., 2006). This work has drawn close attention to what Jim Duncan and the late Denis Cosgrove have argued is the need for the complexities of imperialism and colonialism to be 'unravelled through localized and historically specific accounts' (1995: 127; see also Clayton, 2003).

Locating Colonialism: Scale, the Frontier and the Contact Zone

A key role that geography can play in studies of colonialism is to demonstrate the import of locating analyses in necessarily grounded and differentiated ways. From the 1990s particularly, postcolonial critiques in historical and cultural geography have demonstrated amongst other things the fluidity and hybridity of the multiple *geographies of encounter* in the colonial past, with particular attention paid to the racialized and gendered spaces of colonialism (Blunt and Rose, 1994; Lester, 2001; Blunt and McEwan, 2002). A key conceptual inspiration for much of this work lies in the writings of Homi Bhabha, and particularly his theorization of what he refers to as *third spaces* or *in-between spaces* in which hybridity, ambivalence and mimicry came to characterize colonial cultures, rather than the simple binary opposites of Self and Other, developed by Edward Said (Bhabha, 1994).

The concern with sifting out the social and cultural geographical complexities of colonial encounters has seen a number of key themes explored, including: the agency of the colonized; the nuances of colonialism's in-between spaces; and the co-constitution of colonial cores and peripheries (Kearns, 1984; Lester, 2002). This has also raised important methodological questions of scale and locationality, which historical geographersare well placed to address (Clayton, 2008; Legg, 2010). Core–periphery relationships, for instance, were typically complicated geographically by a range of geopolitical, geoeconomic and symbolic hierarchies and networks. For example, the city of Calcutta operated as core-imperial at the Indian scale but periphery-colonial at the British Empire scale (Legg, 2007).

In addition, geographers have also brought the notion of the frontier into question. In thinking through colonial geographies of encounter, envisioning the past via a frontier lens reinforces notions of geographical boundedness and typically posits self-enclosed regions and ethnicities. This serves to dissuade a reading of the diverse interconnections of cultural contact. As David Wishart reflects, the nature of regions reveals their 'tendency to emphasise differences rather than commonalities, and their limited scope as generalizations' (2004: 305). Conversely, the concept of the *contact zone* has proven particularly useful to historical social and cultural geographical analysis (Pratt, 1992; Routledge, 1997; Morrissey, 2005). Using the idea of the contact zone forms part of a postcolonial dismantling and complicating of the reductive character of colonialism's prioritized geographical knowledges – past and present.

Colonialism and Anti-colonialism

In the general context, colonial expansion in any given geographic setting typically involved key modalities on the ground, subsequent to military, strategic and economic planning (planning that was often more ad hoc than systematic). These include: military conquest and occupation; the establishment of new legal registers to ensure that colonial violence, economic expropriation and dispossession of property were carried out through the law; the mapping of the various lands to be colonized; and the settlement of colonists in new spaces to forge new political, economic and cultural enterprise. All of these colonial modi operandi were not of course always utilized, nor was there any set chronological sequence for colonialism's diverse practices in myriad geographic contexts. However, once established via various mechanisms, colonial order subsequently relied upon a networking of power that facilitated legal, military and political control.

Colonial power of course was resisted all over the world. Anticolonialism typically emerged on a counter-ideological level that initially focused on envisioning the recovery of territory. As Edward Said notes, the political and cultural imagination of anti-colonialism centres on geography:

> For the native, the history of colonial servitude is inaugurated by loss of the locality to the outsider; its geographical identity must thereafter be searched for and somehow restored. Because of the presence of the colonizing outsider, the land is recoverable at first through the imagination. (1993: 271)

From oral histories to radical print cultures, counter-colonial discourses, or what Mary Louise Pratt (1992, 1994) calls *autoethnography*, underpinned and legitimated strategies and practices of resistance (again, the link between discourse and practice). In the nineteenth and twentieth centuries especially, from India to Ireland, from Algeria to Vietnam, that resistance involved a range of practices – from political and economic non-compliance to violence – all mobilized to counter the military and biopolitical control of colonial regimes, with an envisioned endgame of independence.

Practices of political, economic and cultural anti-colonialism ensured the materialization of complex new spaces emerging under the shadow of colonialism. It is important to remember, then, the co-constitution of colonizer and colonized, core and periphery in colonial studies – what Foucault called the 'boomerang effect' (Foucault, 2003: 103). Imperialism in its nineteenth-century form, for example, emerged as 'the ideology of the imperial ruling classes in the very same period that the first substantial freedom movements were developing in the colonies [...] indeed imperialism itself was in part a defensive response to the freedom movements' (Young, 2001: 28). In other words, the colonized worlds informed the nature of the colonial project itself and were a constitutive part in its construction (Lester, 2001).

In many colonial accounts, however, sufficient space is not given to the agency and practices of anti-colonialism (Morrissey, 2004). Indeed, this equates to a key gap in the historiography of colonial and postcolonial studies. Much remains yet to be done in incorporating the historical geographies of decolonization and anti-colonial resistance into our conceptions of the colonial past and present (Clayton, 2008). Given the ongoing Western military, political and ideological failure to comprehend and conceptualize elements of insurgency in Iraq, Afghanistan and elsewhere in the theatre of the war on terrorism, the argument for allowing the historical geographies of anti-colonialism to speak to the present becomes both compelling and urgent.

The Colonial/imperial Present

Despite the end of formal colonialism through the course of the twentieth century, many geographers have alerted us to the imperial power and still hegemonic position of the West and particularly the United States of America in political and economic world affairs (Harvey, 2003; Smith, 2003). In this context, historical geographical work on the twentieth century has addressed a variety of themes, including the operations and spatial strategies of Western interests overseas and the geopolitical and geoeconomic logics of contemporary Western military interventions. Recognizing the echoes of the colonial past in the present has enabled geographers to reveal how global power structures today still mirror the exploitative economic and spatial arrangements of the imperial era. Contemporary geographical work that is historically contextualized has also divulged the enduring import of discourse in the present (particularly the Manichean discursive logics of friend and enemy, and good and evil) and its links to a well-established imperial register of Orientalism.

Various apologiae for the legacies of colonialism, citing its modernizing and civilizing effects, have also emerged in recent years (see particularly the work of Niall Ferguson). Underpinned by a latent imperial nostalgia, such works have reflected on the positive (and apparently bloodless) legacies of colonial endeavours without any critical engagement with the violence, death and destruction of colonialism's civilizing missions, or a recognition of the subaltern contribution to the spaces of colonialism. Much of these accounts can be linked to vociferous calls in the post-9/11 period for a more effective American empire (Ferguson, 2004; Ignatieff, 2004). That *empire* has been seen to be far from benign, of course, frequently exempting itself from international law and the Geneva conventions, and stripping individuals of their most basic human rights and citizenship protections by the invocation of *exceptional* emergency powers that have insidiously become a *norm* in the prosecution of the *war on terror* (Minca, 2005).

Conclusion

In *The Colonial Present*, Derek Gregory traces the specific strategies of contemporary Western interventions in Afghanistan, Palestine and Iraq and illuminates them as 'one more wretched instance of the colonial present'

(2004: 145). Gregory helps us to understand the *war on terror* and its shameful amnesia of past colonialism and indeed perpetuation of an interminable sense of *us* and *them*. The practices and spaces of colonial violence and anti-colonial resistance in the past not only complicate the story of colonialism but also speak specifically to the current moment of global danger and the dichotomy between its representation and materiality, and the frequent effacement of the latter from mainstream Western media.

KEY POINTS

- Colonialism can be thought of as an intrinsically exploitative and dehumanizing system of control that relied upon a networking of legal, military and political power.
- It is important to not only recognize the import of colonial *discourse* in the justification of colonialism but also to see its connections to colonial governmentality and *practice*.
- A key role that geography can play in studies of colonialism is to demonstrate the import of locating analyses in necessarily grounded and differentiated ways, and to this end, the concept of the *contact zone* has been fruitful in sifting out the nuances of colonial encounters.
- Political, economic and cultural practices of anti-colonialism ensured the materialization of complex new spaces emerging under the shadow of colonialism.
- Historical accounts of the practices and spaces of colonial violence and anti-colonial resistance resonate with and illuminate the current moment of global danger in the colonial present.
- Recognizing the echoes of the colonial past in the present has enabled geographers to reveal how global power structures today still mirror the exploitative economic and spatial arrangements of the imperial era.

References

Bhabha, H. (1994) *The Location of Culture*. London: Routledge.

Blunt, A. and McEwan, C. (eds) (2002) *Postcolonial Geographies*. London: Continuum.

Blunt, A. and Rose, G. (eds) (1994) *Writing Women and Space: Colonial and Postcolonial Geographies*. New York: Guilford.

Clayton, D. (2003) 'Critical imperial and colonial geographies', in K. Anderson, M. Domosh, S. Pile and N. Thrift (eds) *Handbook of Cultural Geography*. London: Sage, pp. 354–368.

Clayton, D. (2008) 'Le passé colonial/impérial et l'approche postcoloniale de la géographie Anglophone', in P. Singaravélou (ed.) *L'Empire des Géographes: Géographie, Exploration et Colonisation*, XIXe–XXe Siècle. Paris: Belin, pp. 219–234.

Clayton, D. (2009) 'Colonialism', in D. Gregory, R. Johnston, G. Pratt, M. Watts and S. Whatmore (eds) *The Dictionary of Human Geography* (5th edn). Chichester: Wiley-Blackwell, pp. 94–98.

Duncan, J. and Cosgrove, D. (1995) 'Editorial: colonialism and postcolonialism in the former British Empire', *Ecumene*, 2(2): 127–128.

Ferguson, N. (2002) *Empire: How Britain Made the Modern World*. London: Allen Lane.

Ferguson, N. (2004) *Colossus: The Price of America's Empire*. New York: Penguin.

Foucault, M. (2003) *Society Must be Defended: Lectures at the Collège de France, 1975–1976* (trans. D. Macey). London: Penguin.

Godlewska, A. and Smith, N. (eds) (1994) *Geography and Empire*. Oxford: Blackwell.

Gregory, D. (2004) *The Colonial Present: Afghanistan, Palestine, Iraq*. Oxford: Blackwell.

Harvey, D. (2003) *The New Imperialism*. Oxford: Oxford University Press.

Ignatieff, M. (2004) *The Lesser Evil: Political Ethics in an Age of Terror*. Princeton, NJ: Princeton University Press.

Kearns, G. (1984) 'Closed space and political practice: Frederick Jackson Turner and Halford Mackinder', *Environment and Planning D: Society and Space*, 22: 23–34.

Legg, S. (2007) *Spaces of Colonialism: Delhi's Urban Governmentalities*. Oxford: Blackwell.

Legg, S. (2010) 'Transnationalism and the scalar politics of imperialism', *New Global Studies*, 4(1): 1–7.

Lester, A. (2000) 'Historical geographies of imperialism', in B. Graham and C. Nash (eds) *Modern Historical Geographies*. Harlow: Prentice Hall, pp. 100–120.

Lester, A. (2001) *Imperial Networks: Creating Identities in Nineteenth Century South Africa and Britain*. London: Routledge.

Lester, A. (2002) 'Constructing colonial discourse: Britain, South Africa and the Empire in the nineteenth century', in A. Blunt and C. McEwan (eds) *Postcolonial Geographies*. London: Continuum, pp. 29–45.

Minca, C. (2005) 'The return of the camp', *Progress in Human Geography*, 29(4): 405–412.

Mitchell, T. (1991) *Colonizing Egypt*. Berkeley: University of California Press.

Morrissey, J. (2003) *Negotiating Colonialism*. London: HGRG, Royal Geographical Society.

Morrissey, J. (2004) 'Geography militant: resistance and the essentialisation of identity in colonial Ireland', *Irish Geography*, 37(2): 166–176.

Morrissey, J. (2005) 'Cultural geographies of the contact zone: Gaels, Galls and overlapping territories in late medieval Ireland', *Social and Cultural Geography*, 6(4): 551–566.

Pratt, M.L. (1992) *Imperial Eyes: Travel Writing and Transculturation*. London: Routledge.

Pratt, M.L. (1994) 'Transculturation and autoethnography: Peru, 1615–1980', in F. Barker, P. Hulme and M. Iversen (eds) *Colonial Discourse/Postcolonial Theory*. Manchester: Manchester University Press, pp. 24–46.

Raju, S., Kumar, M.S. and Corbridge, S. (eds) (2006) *Colonial and Post-Colonial Geographies of India*. New Delhi: Sage.
Routledge, P. (1997) 'A spatiality of resistances: theory and practice in Nepal's revolution of 1990', in S. Pile and M. Keith (eds) *Geographies of Resistance*. London: Routledge, pp. 68–86.
Said, E. (1993) *Culture and Imperialism*. New York: Alfred A. Knopf.
Said, E. (2003 [1978]) *Orientalism*. London: Penguin.
Smith, N. (2003) *American Empire: Roosevelt's Geographer and the Prelude to Globalization*. Berkeley: University of California Press.
Wishart, D. (2004) 'Period and region', *Progress in Human Geography*, 28(3): 305–319.
Young, R.J.C. (2001) *Postcolonialism: An Historical Introduction*. Oxford: Blackwell.

FURTHER READING

Blunt, A. and McEwan, C. (eds) (2002) *Postcolonial Geographies*. London: Continuum.
Godlewska, A. and Smith, N. (eds) (1994) *Geography and Empire*. Oxford: Blackwell.
Gregory, D. (2004) *The Colonial Present: Afghanistan, Palestine, Iraq*. Oxford: Blackwell.
Harvey, D. (2003) *The New Imperialism*. Oxford: Oxford University Press.

3 DEVELOPMENT

David Nally

Introduction

In his well-known 'Theses on the Philosophy of History' Walter Benjamin (1992: 249; first published in English in 1968) drew on a painting by Paul Klee, *Angelus Novus* (1920), to forcefully reject any interpretation of history that assumes that humanity is advancing in a discernible, definite and desirable direction. This view Benjamin named 'progress':

> A Klee painting named Angelus Novus shows an angel looking as though he is about to move away from something he is fixedly contemplating. His eyes are staring, his mouth is open, his wings are spread. This is how one pictures the angel of history. His face is turned toward the past. Where we perceive a chain of events, he sees one single catastrophe which keeps piling wreckage upon wreckage and hurls it in front of his feet. The angel would like to stay, awaken the dead and make whole what has been smashed. But a storm is blowing from Paradise; it has got caught in his wings with such violence that the angel can no longer close them. The storm irresistibly propels him into the future to which his back is turned, while the pile of debris before him grows skyward. This storm is what we call progress.

For Benjamin the idea of progress underpinning orthodox conceptions of history fails to acknowledge the horrific destruction and 'piling wreckage' that marks the advance of human civilization. Writing in the wake of the Reichstag fire and Hitler's full assumption of power in Germany, Benjamin was more conscious than most of the perils of fascism and the likelihood of industrial warfare on a terrifying scale. Indeed the 'Theses on the Philosophy of History' was to be the last text Benjamin, a German Jew, ever wrote. Shortly afterwards, on the French–Spanish border, Benjamin took his own life while attempting to flee the Nazis (Arendt, 1968).

'At the gates of the Colosseum and the concentration camp', writes Ronald Wright (2005: 34), 'we have no choice but to abandon hope that civilization is, in itself, a guarantor of moral progress.' The renunciation Wright asks for never came; in fact, within five years of the cessation of Word War II, Harry Truman (1884–1972), the thirty-third President of the United States, would turn the idea of progress into a universal creed applicable to the world at large:

> More than half the world are living in conditions approaching misery. The food is inadequate ... Their poverty is a threat to them and to more prosperous areas ... The United States is pre-eminent among the nations in the development of industrial and scientific techniques ... I believe that we should make available to peace-loving peoples the benefit of our store of technical knowledge in order to help them realize their aspirations for a better life. And, in cooperation with other nations, we should foster capital investment in areas needing development ... The old imperialism – exploitation for foreign profit – has no place in our plans. What we envisage is a program of development based on the concepts of democratic fair-dealing. Greater production is the key to prosperity and peace. And the key to greater production is wider and more vigorous application of modern science and technical knowledge ... To that end we will devote our strength, our resources, and our firmness of resolve. (Cited in Perkins, 1997: 144)

President Truman's inaugural address was an iconic moment. For the first time a whole portion of humanity was characterized as 'needing development' and, at the same time, a positive programme grounded in Western science and technology was put forward as a noble enterprise – differing from the 'old imperialism' (see also Chapter 1) – designed for their collective amelioration. The discourse of progress had transmogrified into a doctrine of development specifying a normative path toward future prosperity. In contrast to Benjamin's reading of *Angelus Novus*, history is here envisaged as a noble tide pushing all boats in one direction.

The fact that the world had just turned from the precipice of all-out war did little to dampen the prescriptive power of the new development mandarin. In fact, under Truman's watch development itself is projected as the only hope of securing enduring peace. So long as 'their poverty is a threat to them and to more prosperous areas', the sitation cannot be left in abeyance. To avoid the 'war of all against all' (*bellum omnium contra omnes*), the secular perdition so memorably described by the English philosopher Thomas Hobbes (1588–1679), societies' collective efforts must be harnessed and channelled into grand productive programmes. If this is done global peace and economic well-being would naturally follow.

Development, in its broadest sense, has of course been an animating vision behind a whole slew of policy prescriptions including the 'Haussmannisation' of Paris (Harvey, 2005); the stadial socialism practised by Stalin (Westad, 2007); the modern metropolis envisioned by Robert Moses; the establishment of the Bretton Woods systems in the mid-twentieth century (Peet, 2009); the 'forced modernization' schemes proposed by Samuel P. Huntington and applied to Vietnam (Berman, 2010); the planned de-colonization of the Third World; the promotion of the Green Revolution (explicitly designed to stave off a 'Red Revolution' – cf. Cullather, 2010); China's one-child policy; forced sterilization campaigns in India; Kim Il-Sung's autarkic and quite disastrous *chuch'e* policies (Haggard and Noland, 2007); the raft of structural adjustment programmes foisted on the global South. One could go on, but suffice to say that 'development', as both vision and practice, has left very little of the world untouched by the grandeur and, arguably, the failure of its promise.

As we can readily see from the different outlooks of Benjamin and Truman, 'development' is a term riddled with contradiction and paradox. This ambiguity is wonderfully exposed in the humorous film *Monty Python's Life of Brian* (1979). In a scene in which the anti-imperial agitators begin to debate the legacy of their nominal oppressors, the Romans, one of the conspirators, played by the actor John Cleese, quickly begins to lose patience: 'All right, all right; but apart from better sanitation and medicine and education and irrigation and public health and roads and a freshwater system and baths and public order – what have the Romans done for us?' The point is a humorous one, of course, levelled at a 'resistance' that wilfully ignored the material benefits brought by the colonizer, but at the same time it is a satire on colonial development, for while the Romans unleashed stunning social changes, embracing new forms of commerce, politics, communication and consumption, they also introduced blood sports, wastefulness, chauvinism and devastation on scales that threatened to subsume the 'civility' they sought to develop spatially through the idea of 'Pax Romana'. What for later Europeans was a cradle of learning, the shining megalopolis of the 'ancient world' ended as a dark necropolis embodying humanity's worst impulses (Mumford, 1961).

For Marshall Berman (2010) – citing Karl Marx – the great caravan of progress generates both promises and perils precisely because 'everything seems pregnant with its contrary' (Marx cited in Berman, 2010: 20). Indeed, as Terry Eagleton (2000: 23) has memorably written, 'dialectical thought arises because it is less and less possible to ignore the fact that civilization, in the very act of realizing some human potentials, also damagingly suppresses others'. For Eagleton and Berman the will to change is always freighted with disaster because the forces we unleash, much like Victor Frankenstein's monster, have the potential to revolt against their putative masters.

It seems important, then, to analyse not only the direction, pace and differing outcomes of social change, but also to think critically about the canonization of the very idea of 'progress' and 'development' within the public sphere. Fredric Jameson's (1981) famous injunction to 'always historicize' implies a need, as Gerry Kearns helpfully reminds us (2006: 125), 'to historicise a particular sense of history'. A similar point is also underscored by Catherine Nash (2000: 27): 'As the representation of the past is so bound up in questions of power and identity, what counts as the past, and what kinds of histories are valued, are often deeply contested.' Thinking in this vein we might ask: how does the notion of development emerge as a universal truth that can simultaneously encompass the social visions espoused by figures as diverse as Chairman Mao, the Dalai Lama and Milton Friedman?

How Development Makes its Object

The elevation of development from a contested belief to the status of metaphysical truth occurred gradually. For the Irish historian J.B. Bury (1861–1927) the idea of progress makes it first appearance as a spiritual concept. It is believers and theologians who appeal most commonly to a state of future felicity – not in this world, it should be said, but in the afterlife (the divine yet-to-be) – and who speak most persuasively and consistently about the 'duty' to care for others as an act of grace and, just as importantly, salvation. In other words, before the notion of progress emerges as an '*idolum saeculi*, the animating and controlling idea of western civilisation' (Bury, 1920: i) it must first convert the idea of felicity in the afterlife into a 'duty to posterity'; and second, it must translate well-known parables such as that of the Good Samaritan (encapsulating a sense of compassion for the fallen) and the legend of St Martin (illuminating the need to share with beggars) into a thoroughly secular compulsion to improve the lives of others.

For social philosopher Marianne Gronemeyer (2010) the elevation of development from a spiritual to a secular concept is linked to this very transposition – what she describes as the 'modernization of the idea of helping'. Gronemeyer contrasts the medieval system of alms-giving (undertaken, she notes, out of 'fearful contemplation' for one's own future in the afterlife) to a thoroughly modern appreciation of assistance as 'other-focused' rather than 'self-centric'. The Victorian architect and social critic Augustus Pugin (1812–1852), in his polemical book *Contrasts* (1836), saw this secularization of charity as an ungodly renunciation of real responsibilities. In this case, the conviction of 'progress' generated a tranche of vocal critics, convinced of the need to return to faith and the communal ethos of a partly or wholly imagined medieval past (Schmiechen, 1988: 292).

Real or imagined, the European Middle Ages served as a measure of modernity. Help is no longer conceived as unconditional assistance granted in order to secure one's own salvation; rather, it is thought of as a calculated act designed to improve a 'deficit' observed in the lives of others (Li, 2007). Furthermore, under the medieval system

> the receipt of alms was neither bound up with humiliating procedures nor in any way made the cause of discrimination. The help was also not educational in relation to the recipient; rather, whatever educational purposes of improvement were connected with help applied much more to the givers. (Gronemeyer, 2010: 57–58)

However, as help becomes a thoroughly secular concept it adopts and internalizes the corollary ideas of progress and improvement, making them an integral aspect of the will to assist. Henceforth help 'implies a view of the cultural and spiritual superiority of the giver. Help still applied to the salvations of souls, but now not to the souls of the givers, but the souls of the recipients' (Gronemeyer, 2010: 60). 'The God of Humanity,' writes Gertrude Himmelfarb (1977: 51), 'proved to be as stern a taskmaster as the God of Christianity'.

The identification of a 'lack' in the lives of the recipients is a central feature of the modern impulse to help: it targets lives and places that exhibit differences that can be constructed as deficiencies (Pogge, 2002). In its contemporary form such help conducts a noble 'struggle against backwardness' (Gronemeyer, 2010: 62) and to this extent the desire to help is also a conscious 'mobilization of the will to break with the past'. In other words, a new concept of historical time is evoked since those in need of succour now appear to exist in a pre-modern state of 'delay' – they wait inert and helpless at the end of an imagined historical queue. Their condition is one of paralysis and help is thought of as a kind of social defibrillator that will 'shock them' out of their enervating stupor and into the present. This modern notion of help – to borrow a phrase of Johannes Fabian (1983) – denies the 'coevalness' of others. The concept of improvement now serves to differentiate, and map, social difference. Hereafter 'higher civilizations' can be measured against 'arrested' or 'petrified civilizations' as Henry George (1839–1897) put it in his classic book *Progress and Poverty* (George, 1966: 186–187).

This is why European travellers in the 'Age of Exploration' (Driver, 2001) persistently described travelling *in* space as travel *through* time. To sojourn in 'darkest Africa' was to bear witness to the pre-history of 'mankind'. This also explains why the 'civilizing mission' and colonial rule marched arm in arm. Native lives were seemingly non-synchronous with the European lives, an argument that gave European rule an aura of respectability that it hardly deserved (see also Chapter 2). It was precisely this will to help – and the corollary notions of progress and improvement – that allowed King Leopold of Belgium to characterize his brutally extractive polices in the Congo as 'humanitarian'. It was decades before anyone in the West challenged Leopold's account because almost everyone believed that European help was the only way to develop backward Africans (for a very moving account see Hochschild, 1999).

It is worth noting here *en passant* that what colonial officials patronizingly termed 'traditional systems' were social forms that exhibited real differences to conventional European society, but were no less modern for that. In fact, as Ronald Wright notes, the striking feature of social evolution is the common achievements of geographically dispersed communities. Wright gives the example of Hernán Cortés (1485–1547) landing in Mexico where 'he found roads, canals, cities, palaces, schools, law courts, markets, irrigation works, kings, priests, temples, peasants, art, music, and books. High civilization, differing in detail but alike in essentials, had evolved independently on both sides of the earth' (Wright, 2005: 51). Similarly the 'Western claim to be the historical depository of ideas on liberty and rights' (Sen, 2006: 93) is not borne out in historical fact. As Amartya Sen points out:

> Emperor Saladin, who fought valiantly for Islam in the Crusades in the twelfth century, could offer, without any contradiction, an honoured place for his Egyptian royal court to Maimonides as that distinguished Jewish philosopher fled an intolerant Europe. When, at the turn of the sixteenth century, the heretic Giordano Bruno was burned at the stake in Campo dei Fiori in Rome, the Great Mughal emperor Akbar (who was born a Muslim and died a Muslim) had just finished, in Agra, his project of legally codifying minority rights, including religious freedom for all. (Sen, 2006: 16)

Other examples could be offered, but the larger argument should be clear: the notion of development as a 'civilizational gift' bestowed by the West – envisaged as the cradle of learning – can only be sustained through a careful forgetting of the achievements of other societies.

Rehabilitating the Poor

Once development shifts from a spiritual to a secular concept – and is firmly associated with measurable levels of social advancement – the concept begins to accrue a scientific status. Newly decorated experts are anointed and tasked with identifying development gaps, social thresholds, key transition phases, population groups deemed to be 'at risk', socio-economic blockages, indicators of progress, and so on; in short, a 'rule of experts' (Mitchell, 2002) is constituted and a new scientific vocabulary for measuring the juggernaut of progress is confirmed.

While contemporary development relies on composite statistics (on income, life expectancy, educational levels, etc.) that enable the production of universal indices (the Human Development Index, for example) as well as the composition of flagship policy reports (such as Food and Agriculture Organization's annual 'State of Food and Agriculture Report'), it is worth recalling that such attempts to capture life in its multiplicity were pioneered by governments seeking to make their populations more amenable to centralised forms of administration (Scott, 1998; see also Chapter 20). A census was needed, for example, to estimate the population in order to determine taxes, but much later the census enumerators delivered standardized forms that recorded each person's name, age, sex and birthplace as well as further particulars about their modes of habitation (the number of rooms in a dwelling, the number of families living there, etc.) and types of employment undertaken by each resident (primary occupation, wages received, etc. – cf. Hannah, 2000). The science of 'demography' and 'political economy' accordingly became vital tools of government and there is no doubt that these changes were driven to a large degree by the progress of industrial capitalism. The careful supervision of the social, indeed the very invention of the 'social', as Michel Foucault (1980) has shown, was predicated on the timely insertion of bodies and whole populations into the machinery of capitalist production (see also Chapter 20). As the hotbed of the Industrial Revolution, Britain proved remarkably innovative in the field of social regulation as schools, poor law workhouses, asylums, hospitals and reformatories were constructed in a bid to maintain or restore social order. In the same way that factories produced 'docile bodies' inured to the hard graft of industrial production, the social institutions of the nineteenth century ensured that the twin imperatives of growth and production insinuated themselves into the knife-and-fork realities of everyday life. A 'great edifice of planning' (Escobar, 2010: 148) was needed to validate and police the work rhythms and normative values that capitalism enshrined in the social order.

As capitalism came to be seen as both a desirable and natural social state – views advanced by the classical economists of the eighteenth and nineteenth centuries (Harcourt, 2011) – social formations and individuals who proved resistant to its work rhythms and values were labelled 'deviant' and subject to remedial interventions designed to transform their conduct. Nowhere is this logic more clearly expressed than in the New Poor Law workhouse. As we have already seen, under the medieval system of alms-giving there was no limiting criteria for gifting help to others and 'consequently', as Gronemeyer (2010: 57) points out, 'there was no distinction, which would later become so indispensible, made between those *unable* to work and those who were *unwilling*'. The English Poor Law of 1834 was the first attempt to ground such a distinction – to give it theoretical form and legal validity. In her marvellous study *The Idea of Poverty*, Gertrude Himmelfarb (1985) argues that the revised Poor Law system brought into being a new category of person – *the pauper* – who was to be cared for and corrected by the state. While the poor might fall out of work and therefore from time to time may struggle to make ends meet (a positive outcome for many Malthusians since it was 'misery and want' that checked population growth amongst the 'lower orders'), it was the pauper, believed to be a social delinquent and constitutionally *unwilling* to work, who was the true object the Poor Law legislation. The Poor Law, through the establishment the workhouse system, aimed to 'dispauperize' the poor; that is, it sought to convert idle and improvident 'good-for-nothings' into sturdy labourers. In a word, it aspired to 'develop' what were previously referred to as the 'undeserving poor' (Driver, 1993; Nally, 2011)

Remaking the World

It is only a short step from correcting the deficiencies of errant groups to the modern notion of development as a kind of a spatio-temporal quest to re-make the world at large (Tyrrell, 2010). In actual fact the line between 'individualized' modes of correction and 'aggregate' improvements targeting whole territories and populations has always been blurred. For example, the Poor Law reforms discussed above were transposed to Ireland where they freely mixed with far more ambitious strategies to modernise Irish society. The architect behind the Irish variant of the New Poor Law, George Nicholls (1781–1865), was quite explicit in his desire to 'transition' Irish peasants from subsistence farming to wage labour on large estates (cited in Nally 2011). High-value commodity production contributed to growth, whereas subsistence farming generated little direct revenue and was condemned as a result.

Of course Nicholls was not the first to attempt what nowadays would be called 'social engineering'; the historical geographer H.C. Darby (1973) described the period from 1600–1800 as 'the age of the improver'. During this two-hundred year stretch the 'waste' and marshlands of England were drained, heathlands were reclaimed, grounds held in common were enclosed and privatized, new breeds of cattle and sheep were introduced, experimental forms of landscaping and farming (leaving certain arable lands under grass, the ploughing up of pasture lands, etc.) were put into practice. The spirit of progress filtered through the English countryside.

No, what is perhaps new about Nicholls's plans for improvement is the belief that the mode for modernization in England can be universally extended. This nurturing creed marks the beginning of the period of 'liberal empire' (Duffield and Hewitt, 2009; Lester, 2012; Skinner and Lester, 2012). Gronemeyer (2010: 57) describes this as a 'metamorphosis from a colonialism that "takes" to one that supposedly "gives"'. The promise of progress lends reformers like Nicholls a legitimising discourse of extraordinary power. Under the aegis of development impositions of all kinds can be re-presented as so many gifts to the natives (Olund, 2002; Lester, 2013).

According to Marshall Berman (2010) development has a habit of turning quintessentially modern formations into old-fashioned and obsolete relics – so many encumbrances to be targeted by further rounds of 'improvement' and 'modernization'. It should be said that this disillusionment and weariness towards the present is not a unique feature of *capitalist development* (though Harvey is right to conclude that the historical geography of capitalism is defined by the tendency to construct one landscape at one point in time only to turn around and destroy it at a later date in order to construct another). Indeed, the 'ideological geopolitics' (Agnew and Corbridge, 1995) of the Cold War era pitched a 'Marxist modernity', built on the promise of a classless society, against modern capitalism with its promise of democracy, open competition and civil liberties. At their core, both ideologies laid claim to the 'mantle of modernity'. The Soviets and the Americans believed that they were acting in the best interests of humanity and over the course of the Cold War they proved themselves more than willing to project those views outwards – forcefully if necessary – onto the global sphere in a manner that for Odd Arne Westad (2007) recalls the hubris and thinly veiled jingoism of liberal imperialists (Smith, 2005). The 'strategic hamlet program' operated by the United States in Vietnam and the collectivization schemes established by the Derg in Ethiopia (under Soviet tutelage) are salient reminders of the costs involved in pressing social changes on supposedly backward populations.

Conclusion

Oscar Wilde (1854–1900) said of utopias and progress:

> A map of the world that does not include Utopia is not worth even glancing at, for it leaves out the one country at which Humanity is always landing. And when Humanity lands there, it looks out, and, seeing a better country, sets sail. Progress is the realisation of Utopias. (Wilde, 1910: 27)

To extend the martime metaphor, but add to it a slightly darker resonance, we might say that 'progress' is akin to the sweet songs employed by the mythic Sirens to tempt the lost mariner Odysseus; it beguiles its listeners, promising fulfilment only to dash those hopes upon jagged rocks. For this very reason Wolfgang Sachs (2010: xviii) suggestively argues that 'it is not the failure of development which has to be feared, but its success'.

Is what geographer Tony Weis (2007: 15) in a different context calls 'planetary convergence' actually a desirable outcome? When experts talk about developing low-income countries – or more hubristically 'Africa' – what terminal stage are they envisaging? Scientists tell us that if the entire world was to consume the same amount of resources as the average American we would require four or five additional planets – and yet Western values and norms tend to saturate much of development-speak today (Tyrrell, 2010).

These and related questions have been posed by scholars critical of development discourse. Some such as Arturo Escobar (1995) have proclaimed that the concept of development is 'dead' and consequently we have now entered a 'post-development' era (Rahnema, 1997; Kapoor, 2008). Certainly development as a monolithic discourse has been knocked by these criticisms (cf. Corbridge, 1998; McEwan, 2001), but it still seems far too premature to anoint the corpse. Turn on the radio, open a newspaper, download a podcast or switch on the television and it will not be long before you hear some scholar, politician or expert declare that 'development aid is dead' and micro-credit loans are in fact a better answer to global poverty. Equally when you hear scientists proclaim that climate change means we need to 'transition' to a 'weightless society' or experiment with 'geo-engineering' (Foster et al., 2010); when food experts call for policies to curb population growth and endorse genetically modified organisms (GMOs) to boost agricultural production; when technocrats encourage us to embrace cyborg identities and simulation cultures or urge us to 're-world' to cyberspace; when economists and business analysts promote futures trading as a means to revive the global financial system; when the Chinese government begins to buy up large swathes of land in the global South, whilst appealing to the benefits these investment will bring to local populations (Nally, 2012) – in each case a claim is made about a positive trajectory of change that sanctions the suite of interventions being proposed (Watts, 2003). It makes sense in this context to speak of the emergence of 'zombie development' in the same way that Chris Harman (2009) has described the emergence of 'zombie capitalism'. Development might be declared 'dead' but nevertheless it is still upright and walking!

KEY POINTS

- The idea of development may be traced through the religious obligation to do 'good deeds' in order to secure one's salvation in the afterlife. As the concept is secularized it becomes less about the spiritual redemption of 'givers' and more about the worldly salvation of the 'recipients'.
- At its core the concept of development internalizes what Wolfgang Sachs (2010: x) calls 'chronopolitics'; that is a need for 'backward peoples' to catch up with the 'pacemakers [of history] who are supposed to represent the forefront of social evolution'.

- The emergence of industrial capitalism brought with it a host of disciplinary powers whose object was to retrofit 'deviant' bodies for capitalist production. The age of empire channelled these ideals overseas where the 'will to improve' became an alibi for a 'humanitarian' mode of empire.
- Contemporary development, though now rooted in the language of technical and scientific innovation, still tends to formulate 'social evolution' in the quintessentially Western term.

References

Agnew, J. and Corbridge S. (1995) *Mastering Space: Hegemony, Territory and International Political Economy*. London: Routledge.

Arendt, H. (1968) *Men in Dark Times*. New York: HBJ Book.

Benjamin, W. (1992) *Illuminations* (edited with an introduction by Hannah Arendt; translated by Harry Zohn). London: Fontana Press.

Berman, M. (2010) *All That Is Solid Melts into Air: The Experience of Modernity*. London: Verso.

Bury, J.B. (1920) *The Idea of Progress: An Inquiry into its Origin and Growth*. London: Macmillan.

Corbridge, S. (1998) '"Beneath the pavement only soil": the poverty of post-development', *Journal of Development Studies*, 34(6): 138–148.

Cullather, N. (2010) *The Hungry World: America's Cold War Battle Against Poverty in Asia*. Cambridge, MA: Harvard University Press.

Darby, H.C. (1973) 'The age of the improver: 1600–1800', in H.C. Darby (ed.) *A New Historical Geography of England*. Cambridge: Cambridge University Press, pp. 302–388.

Driver, F. (1993) *Power and Pauperism: The Workhouse System, 1834–1884*. Cambridge: Cambridge University Press.

Driver, F. (2001) *Geography Militant: Cultures of Exploration and Empire*. London: Blackwell.

Duffield, M. and Hewitt, V. (eds) (2009) *Empire, Development and Colonialism: The Past in the Present*. Oxford: James Currey.

Eagleton, T. (2000) *The Idea of Culture*. Oxford: Blackwell.

Escobar, A. (1995) *Encountering Development: The Making of the Third World*. Princeton, NJ: Princeton University Press.

Escobar, A. (2010) 'Planning' in W. Sachs (ed.) *The Development Dictionary: A Guide to Knowledge as Power* (2nd edn). London: Zed Books, pp. 145–160.

Fabian, J. (1983) *Time and the Other: How Anthropology Makes it Object*. New York: Columbia University Press.

Foucault, M. (1980) *The History of Sexuality. Vol. 1, An Introduction*. New York: Vintage Books.

Foster, J.B., Clark, B. and York, R. (2010) *The Ecological Rift: Capitalism's War on the Earth*. New York: Monthly Review Press.

George, H. (1966) *Progress and Poverty*. London: The Howarth Press.
Gronemeyer, M. (2010) 'Helping', in W. Sachs (ed.) *The Development Dictionary: A Guide to Knowledge as Power* (2nd edn). London: Zed Books, pp. 55–73.
Haggard, S. and Noland, M. (2007) *Famine in North Korea: Markets, Aid, and Reform*. New York: Columbia University Press.
Hannah, M. (2000) *Governmentality and the Mastery of Territory in Nineteenth-Century America*. Cambridge: Cambridge University Press.
Harman, C. (2009) *Zombie Capitalism*. London: Bookmarks.
Harcourt, B.E. (2011) *The Illusion of the Free Markets: Punishment and the Myth of Natural Order*. Cambridge, MA: Harvard University Press.
Harvey, D. (2005) *Paris, Capital of Modernity* (2nd edn). London: Routledge.
Himmelfarb, G. (1977) 'The Age of Philanthropy', *The Wilson Quarterly*, 21(2): 48–55.
Himmelfarb, G. (1985) *The Idea of Poverty: England in the Early Industrial Age*. New York: Vintage Books.
Hochschild, A. (1999) *King Leopold's Ghost: A Story of Greed, Terror and Heroism in Colonial Africa*. Boston: Houghton Mifflin Company.
Jameson, F. (1981) *The Political Unconscious: Narrative as a Socially Symbolic Act*. Ithaca, NY: Cornell University Press.
Kapoor, I. (2008) *The Postcolonial Politics of Development*. New York: Routledge.
Kearns, G. (2006) 'The spatial politics of James Joyce', *New Formations*, 57(1): 107–125.
Lester, A. (2012) 'Humanism, race and the colonial frontier', *Transactions of the Institute of British Geographers*, 37(1): 132–148.
Lester, A. (2013) 'Benevolent empire? Protecting indigenous peoples in British Australasia', in R. Crane, A. Johnston, A. and C. Vijayasree (eds) *Empire Calling: Administering Colonial Australasia and India*. New Delhi: Cambridge University Press, pp. 3–23.

Li, T. M. (2007) *The Will to Improve: Governmentality, Development, and the Practice of Politics*. Durham, NC: Duke University Press.
McEwan, C. (2001) 'Postcolonialism, feminism and development: intersections and dilemmas', *Progress in Development Studies*, 1(2): 93–111.
Mitchell, T. (2002) *Rule of Experts: Egypt, Techno-politics, Modernity*. Berkeley: University of California Press.
Monty Python's Life of Brian (1979) Dir. Terry Jones, Cinema International Corp.
Mumford, L. (1961) *The City in History*. Harmondsworth: Penguin.
Nally, D. (2011) *Human Encumbrances: Political Violence and the Great Irish Famine*. Notre Dame, IN: University of Notre Dame Press.
Nally, D. (2012) 'Trajectories of development, modalities of enclosure: land grabs and the struggle over geography', in P.J. Duffy and W. Nolan (eds) *At the Anvil: Essays in Honour of William J. Smyth*. Dublin: Geography Publications, pp. 653–676.
Nash, C. (2000) 'Historical geographies of modernity', in B. Graham and C. Nash (eds) *Modern Historical Geographies*. London: Pearson, pp. 13–37.
Olund, E. (2002) 'From savage space to governable space: the extension of the United States judicial sovereignty over Indian country in the nineteenth century', *Cultural Geographies*, 9(2): 129–157.
Peet, R. (2009) *Unholy Trinity: The IMF, World Bank and WTO* (2nd edn). London: Zed Books.
Perkins, J. (1997) *Geopolitics and the Green Revolution: Wheat, Genes, and the Cold War*. Oxford: Oxford University Press,

Pogge, T. (2002) *World Poverty and Human Rights: Cosmopolitan Responsibilities and Reforms*. Polity: Cambridge.
Pugin, A. (1836) *Contrasts: Or, A Parallel Between the Noble Edifices of the Fourteenth and Fifteenth Centuries and Similar Buildings of the Present Day. Shewing the Present Decay of Taste*. London: James Moyes.
Rahnema, M. (ed.) (1997) *The Post-Development Reader*. London: Zed.
Sachs, W. (ed.) (2010) *The Development Dictionary: A Guide to Knowledge as Power* (2nd edn). London: Zed Books.
Schmiechen, J.A. (1988) 'The Victorians, the historians, and the idea of modernism', *American Historical Review*, 93(2): 287–316.
Scott, J. (1998) *Seeing Like a State: How Certain Schemes to Improve the Human Condition Have Failed*. New Haven, CT: Yale University Press.
Sen, A. (2006) *Identity and Violence: The Illusion of Destiny*. New York: W.W. Norton & Company.
Skinner, R. and Lester, A. (2012) 'Humanitarianism and empire: new research agendas', *Journal of Imperial and Commonwealth History*, 40 (5): 729–747.
Smith, N. (2005) *The Endgame of Globalization*. New York: Routledge.
Tyrrell, I. (2010). *Reforming the World: The Creation of America's Moral Empire*. Princeton: Princeton University Press.
Watts, M. (2003) 'Development and governmentality', *Singapore Journal of Tropical Geography*, 24(1): 6–34.
Weis, T. (2007) *The Global Food Economy: The Battle for the Future of Farming*. New York: Zed Books.
Westad, O.A. (2007) *The Global Cold War: Third World Interventions and the Making of Our Times*. Cambridge: Cambridge University Press.
Wilde, O. (1910) *The Soul of Man Under Socialism*. Boston: John W. Luce.
Wright, R. (2005) *A Short History of Progress*. New York: Carroll & Graf Publishers.

FURTHER READING

Berman, M. (2010) *All That Is Solid Melts into Air: The Experience of Modernity*. London: Verso.
Escobar, A. (1995) *Encountering Development: The Making of the Third World*. Princeton, NJ: Princeton University Press.
Sachs, W. (ed.) (2010) *The Development Dictionary: A Guide to Knowledge as Power* (2nd edn). London: Zed Books.
Skinner, R. and Lester, A. (2012) 'Humanitarianism and empire: new research agendas', *Journal of Imperial and Commonwealth History*, 40(5): 729–747.

Section 2 Nation-building and Geopolitics

4 TERRITORY AND PLACE

Yvonne Whelan

Introduction: Mapping the Terrain of Territory

> Terra means land, earth, nourishment, sustenance; it conveys the sense of a sustaining medium, solid, fading off into indefiniteness. But the form of the word, the OED says, suggests that it derives from *terrere*, meaning to frighten, to terrorize. And *Territorium* is a "place from which people are warned." Perhaps these two contending derivations continue to occupy territory today. To occupy a territory is to receive sustenance and to exercise violence. Territory is land occupied by violence. (Connolly, 1996: 144)

Our daily news bulletins are often dominated by stories about territorial struggles of one kind or another, whether it is disputed national boundaries in the global arena, or a local neighbourhood dispute as Connolly alludes to in the quote above. We use the word 'territory' to refer to bounded places, for example the land that is claimed by a particular country. As Storey writes, 'territory refers to a portion of geographic space which is claimed or occupied by a person or group of persons or by an institution. It is, therefore, an area of bounded space, sometimes quite literally surrounded by a wall, fence or barrier of some sort' (2001: 1). The 'claiming' and 'occupation' of space is significant here, for territory in many ways is about claiming ownership, taking over and occupying a particular terrain. Unlike the more benign concept that is 'place', with its connotations of space made meaningful by human habitation, territory and territoriality are suggestive of more malign forces at work. The bounded social spaces that go hand-in-hand with territory are invariably a result of the adoption of strategies of territoriality, whereby people, groups or organizations exercise power and control over a particular place and its component parts (Sack,

1986). The allocation of people or social groups to distinct areas or territories, invariably separated by borders or boundaries, is at the heart of human territoriality (Grosby, 2005).

Various geopolitical rivalries and the attendant pursuit of political power have ensured that numerous territorial legacies have been carved out in landscapes across the globe. Colonial and empire-building projects in the nineteenth century, for example, created a marked territorial legacy throughout parts of Africa, Asia and the Americas, just as in later years the Cold War would generate another set of territorial legacies throughout Eastern Europe. Perhaps one of the most obvious forms of territorial organization is the division of the world's territory into distinct political states or nation-states. As a form of spatial governance, the nation-state is underpinned by a territorial ideology that links people with places and in such contexts landscape and territorial imagery are hugely important for the evocation of the nation (Gellner, 1983; Anderson, 1990). But territory and territoriality also operate at the micro-level, a scale far removed from grand empire-building projects or state formation. Everyday spaces are also subject to territorial control and strategies of territoriality, from the domain of the city through to the personal space that exists around the individual. Within urban areas, for example, clear spatial divisions can be seen to exist with 'fault lines dividing zones in terms of affluence, class or ethnicity' (Storey, 2001: 5). Territory, therefore, is implicated in a wide range of social relations, from the interpersonal to the international and many different forms and expressions of power are implicated in its constitution and maintenance (Delaney, 2005).

Over the course of generations a large amount of knowledge has been produced about the forms, functions and processes associated with territoriality and boundaries, from the macro through to the micro level. For geographers, territory has long been a key concern, overlapping and intertwining with concepts like place, space, power and the cultural landscape (Sack, 1986; Agnew, 1994, 1997; Cresswell, 2004). Almost exclusively the preserve of political geographers prior to the 1970s, territory and the nation-state constituted a key strand of inquiry, from the work of Friedrich Ratzel on territory and boundaries in the mid-nineteenth century, through to Richard Hartshorne's 'disinterested analysis of territory or "politically organised area"' in the mid-1950s (Delaney, 2005: 41). Geopolitics and territorial organization remained a strong focus of inquiry throughout the twentieth century with many scholars exploring political boundaries and the territorial tensions surrounding annexation and partition (Johnston, 2001). Others examined the workings of territory *within* the state as well as in various non-state contexts (Ley, 1983). As the discipline of human geography began to change in the late 1970s, however, so too did the analysis and treatment of territory. Interdisciplinary approaches and new theoretical insights had a particular impact on the treatment of territoriality. As Delaney argues:

> scholars from various "home" disciplines began to increasingly explore ways of escaping the territoriality of territory through a range of interdisciplinary projects. Drawing on an expanded array of theoretical resources such as post-structuralism, post-modernism, political economy, and feminism, these writers began to express a more explicit reflexivity regarding the production of knowledge and the ways in which established disciplinary perspectives both focus attention on some aspects of territoriality and limit the scope of inquiry. There has been an increased awareness of the relationship between knowledge (representations of territory) and power. (2005: 52)

One of the seminal texts to explore issues of human territoriality was written in 1986 by the geographer Robert Sack entitled *Human Territoriality: Its Theory and History*. Situating his discussion of territory within the sub-fields of social and historical geography, rather than political geography, and drawing insights from a range of interdisciplinary perspectives, Sack foregrounds the relationship between territoriality and power, arguing that territoriality is the primary expression of social power and can be defined as 'the attempt by an individual or group to affect, influence, or control people, phenomena and relationships, by delimiting and asserting control over a geographic area' (1986: 19). He ranges over a wide variety of spatial scales, from state formation through to territoriality in the domestic sphere of the home, arguing all the while that people are driven to behave in a territorial manner not because it is instinctive but rather because of particular sets of political, social or cultural circumstances. Thus, he places human territoriality 'entirely within the control of human motivations and goals' (Sack, 1986: 21). Since the publication of *Human Territoriality* the development of a more critical geopolitics has led to a destabilizing of notions of territory and a reconceptualization in contemporary geography. Informed by the work of Lefebvre (1991) and Soja (1989), territory has increasingly come to be conceptualized 'not as a passive grid of lines and space but rather as inextricably implicated one way or another in virtually all aspects of human social action, being, consciousness and experience' (Delaney,

2005: 60). Scholars have drawn attention to how territoriality operates among the more marginalized and oppressed, and have examined the spatiality of interpersonal relationships, as well the ways in which social constructs such as class, gender and ethnicity influence territory and territorialization (Sibley, 1995; McDowell and Sharp, 1997; Domosh and Seager, 2001). The remainder of the chapter explores some of the ways in which territory operates as a component of group identity and how it can be claimed, controlled but also contested and resisted at the micro level. Using the example of territorial conflict in Northern Ireland, I want to probe further some of the ways in which territorial behaviour manifests, as well as how territorial markers and boundaries exert control over public space.

Contested Territory and the Urban Landscape

The territorial conflict in the north of Ireland is longstanding and bitterly contested. It stretches back to the sixteenth century when Protestant settlers came to live in close proximity to the Catholic Irish who were in turn relegated to more marginal locations. Over the generations that followed the Ulster Plantation, this territory was shared uneasily by rival ethnic groups, with differing religious practices, cultural values and political allegiances. The partitioning of Ireland in 1921 established the Irish Free State (which would later become the Republic of Ireland in 1949), while six counties of the province of Ulster remained part of the UK. This paved the way for a heightened territorial conflict that would last into the twenty-first century. The new state of Northern Ireland, with its own parliament and a large degree of autonomy from Westminster, was marked by high levels of demographic and social segregation. Tension flared up in the 1960s when the civil rights movement began to campaign for equal access to housing, employment and political power. Since then, over 3,600 people have died out of a relatively small population of 1.6 million people. The victims comprised civilians, members of the security forces, and members of the diverse range of loyalist and republican paramilitary organizations.

The Troubles, as the conflict is often referred to, have been characterized by a whole range of territorial struggles that have been played out across this contested terrain. The territory of the six counties as a whole has been at the heart of the conflict, along with the border territory that separates these counties from the remaining counties of Ulster and the Republic of Ireland. These larger-scale territorial disputes are in many ways sustained by a wide variety of what we might term micro-scale symbolic strategies across the cultural landscapes of the province. Territorial signifiers have been employed by both communities in order to represent and reaffirm group identity, as well as to create clearly demarcated boundaries between communities (Buckley, 1998; Yiftachel and Ghanem, 2004). In the contested space of Northern Ireland 'hegemonic struggles for territory, recognition and constitutional status are frequently constituted in appeals to "history", "culture" and "tradition"' (Wilson and Stapleton, 2005: 634). So, urban and rural landscapes have been marked by a large number of unofficial memorials erected by republican organizations, chiefly Sinn Féin and the Irish Republican Army (IRA), and by the more fragmented loyalist paramilitary groups, notably the Ulster Defence Association (UDA) and Ulster Volunteer Force (UVF).

In the city of Derry, for example, artists have created a trail of gable wall murals that graphically re-create key events of the Troubles (Kelly, 2001). These Bogside murals frame the commemorative centrepiece, a memorial cross at 'Free Derry' corner, erected in 1974 in memory of the victims of 'Bloody Sunday' (Figure 4.1). Close by, a granite monument built in 2001 to commemorate the 25th anniversary of the Maze Prison hunger strike lists the names of each of those who took part, together with plaques honouring hunger strikers who died in the 1920s. This commemorative tableau is completed by a ten-foot high statue of a paramilitary fighter that was unveiled in March 2000 at the graves of two 1981 hunger strikers in the nearby Creggan cemetery. Those who commissioned the monument saw it as 'a fitting tribute to members of the Irish National Liberation Army and Irish Republican Socialist Party from counties Derry and Tyrone who gave their lives during the latest phase of the war against the British establishment in Ireland'. These examples of 'community remembrancing' serve as subversive icons of identity that transform 'neutral' spaces into ideologically charged sites of symbolic struggle. They also mark out territory and as improvised, piecemeal memorials they represent strategic displays of resistance through remembrance. For republicans, events such as Derry's 'Bloody Sunday' which took place in 1972 and the 1981 IRA hunger strike are woven into the iconography of the landscape, whereas loyalists frequently draw on the imagery and rhetoric of the Battle of the Somme in order to align their recent dead with those of World

Figure 4.1 Memorial spaces of the Bogside, Derry (Photo: Yvonne Whelan)

War I (Graham and Shirlow, 2002). While such unofficial memorial landscapes sustain and legitimate a bank of shared memories, they are also invariably sectarian spaces that serve to divide rather than unite by marking territorial boundaries and shaping situated identities in contradistinction to the ethnic other. Further, as sites of 'remembrance resistance', monuments, murals and even graffiti represent internal divisions *between* and *within* communities. Thus, the splintering of loyalism has been marked by attempts on the part of the UVF and the Ulster Freedom Fighters (UFF) to control and claim social space on the Shankill Road, Belfast, where intracommunal animosities can be read via the iconography of adjacent wall murals.

Amidst this highly fraught political context, territory in parts of the province came to be defined by a whole range of human rituals and symbols. Paramilitary groupings found political wall murals to be a particularly potent and relatively inexpensive means of staging strategic forms of remembrance, as well as serving as tangible expressions of territorial control. Although potentially more ephemeral than monuments, murals effectively claim space at a community level and often mobilise local populations in support of particular political initiatives. They also serve as a means of reinforcing division and cultural difference and as such 'are not just a backdrop to politics, but a dynamic part of the political process' (Rolston, 2003: 14). The tradition of mural painting in Northern Ireland stretches back to 1908 when a group of loyalist artisans set about painting large outdoor murals each 12 July as part of the annual commemoration of the Battle of the Boyne when the Protestant King William III defeated the Catholic King James II. This, in turn, became the most popular image used by loyalists in working-class districts. Gradually, murals that were once painted annually for the 12 July celebrations became more permanent fixtures of the cultural landscape and expressed in highly visual terms an increasing sectarianism of place. As Jarman writes, loyalist wall murals effectively marked out streets as being loyal and Protestant:

> Streets could now declare their faith throughout the year. They were no longer simply rows of houses, but terraces of Protestant houses. The permanent displays visually confirmed a status above and beyond mere function. While this may have been a relatively minor step in areas which were already recognised as staunchly Protestant, and much of working-class Belfast was highly segregated, such displays helped to make explicit the fact of residential segregation. (1998: 84)

During the 1980s and into the 1990s wall murals remained an enormously popular means of claiming space, defining territory, strengthening group identity and mobilizing the local population in support of different political causes and ideological struggles. Republicans began to make use of wall murals as a kind of spatial mnemonic, especially during the hunger-strike period of the early 1980s. Numerous murals were painted in nationalist spaces of Belfast, some streets were renamed in Irish, while green, white and gold kerbstones asserted 'a permanent and visible, political and cultural dominance over the area' (Jarman, 1998: 86). Thus, mural painting, which had once been prohibited in nationalist areas, became an important means of expression. In subsequent years these murals often made a particular point of representing specific aspects of Irish culture and history, thereby underscoring the political allegiance to Ireland and resistance to the British state (Rolston, 1991, 1992, 1995). A resurgence in mural painting in loyalist areas followed, as both communities sought to advance their respective causes. These murals often tended to draw on an altogether different and more overtly militaristic set of images which expressed fervent loyalty to the British state and bitter opposition to the cause of a united Ireland (Figure 4.2).

Figure 4.2 The 'Grim Reaper' wall mural, Belfast (Photo: Yvonne Whelan)

Conclusion

The case study of the hotly contested and deeply segregated nature of territory in the north of Ireland captures some of the issues that attend human territoriality more broadly, especially in zones of conflict. As a symbolic strategy, the painting of wall murals is just one means of appropriating the cultural landscape in order to bolster territorial control. We could also ponder the ubiquitous 'peace-lines' that separate communities in parts of Belfast, for example, or the painting of kerbstones in some loyalist and republican areas. As the political context in Northern Ireland stabilizes, however, the territorial markers referred to above are beginning to take on an altogether different set of functions and purposes. What were once the primary symbolic signifiers of conflict and spaces which reinforced division and fear are today taking on new meanings in a 'post-conflict', peace process, context. In July 2006, for example, the government announced a £3.3 million scheme to replace paramilitary wall murals as part of a communities action plan scheme. The 'Re-Imaging Communities Programme' is one of 62 actions included in the Renewing Communities Action Plan which has set about engaging with local people and their communities in order to find ways of replacing divisive murals and emblems with more positive imagery. The aim is to create new murals and public art which 'will transform parks, housing estates and built-up areas across Northern Ireland, celebrating the aspirations of the whole community and helping people feel part of their own local community' (Arts Council of Northern Ireland, July 2006). So, what were once landscapes of fear, comprising numerous territorial divisors, are very gradually being rewritten as shared spaces of heritage.

KEY POINTS

- For geographers, territory has long been a key concern, overlapping and intertwining with concepts like place, space, power and the cultural landscape. Territory refers to bounded places, for example the land that is claimed or occupied by a particular country, institution or person
- The allocation of people or social groups to distinct areas or territories, invariably separated by borders or boundaries is at the heart of human territoriality, a term used to describe the strategies used by people, groups or organisations to exercise power and control over a particular place and its component parts.
- Territory has increasingly come to be conceptualized not as a passive grid of lines and space but rather as inextricably implicated one way or another in virtually all aspects of human social interaction, from state formation through to the personal space that surrounds an individual.
- Over the course of generations a large amount of knowledge has been produced about the forms, functions and processes associated with territoriality and boundaries, from the macro through to the micro level.
- The case study of Northern Ireland captures in microcosm some of the issues that attend human territoriality more broadly, and especially in zones of conflict whereby symbolic signifiers, such as wall murals, take on particular territorial significance.

References

Arts Council of Northern Ireland (2006) *Reimagining Communities Programme: Creating a Welcoming Environment for Everyone*. Belfast: Arts Council.

Agnew, J.A. (1994) 'The territorial trap: the geographical assumptions of international relations theory', *Review of International Political Economy*, 1(1): 53–80.

Agnew, J.A. (1997) *Political Geography: A Reader*. London: Arnold.

Anderson, B. (1990) *Imagined Communities*. London: Verso.

Buckley, A. (1998) 'Introduction: daring us to laugh: creativity and power in Northern Irish symbols', in A. Buckley (ed.) *Symbols in Northern Ireland*. Belfast: Institute of Irish Studies, pp. 1–22.

Connolly, W. (1996) 'Tocqueville, territory and violence', in M. Shapiro and H. Alker (eds) *Challenging Boundaries: Global Flows, Territorial Identities*. Minneapolis: University of Minnesota Press, pp. 141–164.

Cresswell, T. (2004) *Place*. Oxford: Blackwell.

Delaney, D. (2005) *Territory*. Oxford: Blackwell.

Domosh, M. and Seager, J. (2001) *Putting Women in Place: Feminist Geographers Make Sense of the World*. New York: Guilford Press.

Gellner, E. (1983) *Nations and Nationalism*. Oxford: Blackwell.

Graham, B. and Shirlow, P. (2002) 'The Battle of the Somme in Ulster memory and identity', *Political Geography*, 21(7): 881–904.

Grosby, S. (2005) 'Territoriality: the transcendental primordial feature of modern societies', *Nations and Nationalism*, 1(2): 143–162.

Jarman, N. (1998) 'Painting landscape: the place of murals in the symbolic construction of urban space', in A. Buckley (ed.) *Symbols in Northern Ireland*. Belfast: Institute of Irish Studies, pp. 81–98.

Johnston, R. (2001) 'Out of the "moribund backwater": territory and territoriality in political geography', *Political Geography*, 20: 677–693.

Kelly, T. (2001) *Murals: the Bogside Artists*. Derry: Guildhall Press.

Lefebvre, H. (1991) *The Production of Space*. London: Blackwell.

Ley, D. (1983) *A Social Geography of the City*. New York: Harper & Row.

McDowell, L. and Sharp, J. (eds) (1997) *Space, Gender and Knowledge: Feminist Readings*. London: Arnold.

Rolston, B. (1991) *Politics and Painting: Murals and Conflict in Northern Ireland*. Cranbury, NJ: Associated University Presses.

Rolston, B. (1992) *Drawing Support: Murals in the North of Ireland*. Belfast: Beyond the Pale Publications.

Rolston, B. (1995) *Drawing Support 2: Murals of War and Peace*. Belfast: Beyond the Pale Publications.

Rolston, B. (2003) 'Changing the political landscape: murals and transition in Northern Ireland', *Irish Studies Review*, 11(1): 3–16.

Sack, R. (1986) *Human Territoriality: Its Theory and History*. Cambridge: Cambridge University Press.

Sibley, D. (1995) *Geographies of Exclusion. Society and Difference in the West*. London: Routledge.

Soja, E. (1989) *Postmodern Geographies. The Reassertion of Space in Critical Social Theory*. London: Verso Press.

Storey, D. (2001) *Territory: The Claiming of Space*. Harlow: Pearson Education.

Wilson, J. and Stapleton, K. (2005) 'Voices of commemoration: the discourse of celebration and confrontation in Northern Ireland', *Text and Talk*, 25(5): 633–644.

Yiftachel, O. and Ghanem, A. (2004) 'Understanding "ethnocratic" regimes: the politics of seizing contested territories', *Nations and Nationalism*, 23(6): 647–676.

FURTHER READING

Delaney, D. (2005) *Territory*. Oxford: Blackwell.
Jarman, N. (1998) 'Painting landscape: the place of murals in the symbolic construction of urban space', in A. Buckley (ed.) *Symbols in Northern Ireland*. Belfast: Institute of Irish Studies, pp. 81–98.
Johnston, R. (2001) 'Out of the "moribund backwater": territory and territoriality in political geography', *Political Geography*, 20: 677–93.
Sack, R. (1986) *Human Territoriality: Its Theory and History*. Cambridge: Cambridge University Press.

5 IDENTITY AND THE NATION

John Morrissey

Introduction

Historical geographers have long been concerned with matters of identity. Contexts from colonialism to capitalism have been variously examined, with studies spanning a range of issues, including class, power and resistance. Elsewhere in this volume, questions of identity are engaged around these and other axes of historical inquiry such as race and gender. This chapter, however, takes a particular focus on the historical geographies of *nationhood*, and using various examples explores how tropes of *national identity* were over time constructed and reproduced spatially. It examines too the theoretical concepts of sameness and difference upon which all national identities have been forged, and in outlining how diverse senses of identity networked in myriad historical geographical contexts the chapter complicates the notion of national identity by demonstrating how essentialist narratives of the nation limit our understanding of the complexities of the worlds of the past.

Constructing the Nation

Even the world's oldest nations do not have long histories as states. While the exact origins of the phenomenon of nationalism are contested, a common misconception is that sovereign nations extend back to ancient times (Hastings, 1997). Most states, however, date to the nineteenth and twentieth centuries when nations across the world were embroiled in prolonged nationalist struggles against the main imperial powers. Others only gained independence more recently, as seen with the break-up of the Soviet Union and Yugoslavia in the 1990s. And although most nations today could be argued to be in some respects post-nationalist and globalized, such a contention must be tempered with the reality of ongoing fights for independent statehood in various regions of the world that claim historical nationhood, from the Basque Country to Palestine, from South Ossetia to Kurdistan. Nationalism, in other words, still matters (Castells, 1997).

If most states are relatively new, how have they been constructed and how can historical geographers interrogate the process of nation-building? To begin with, a key concept in identity studies more generally is that all identities are *socially constructed.* As Cindi Katz (2003: 262) argues, senses of identity do not rest 'upon some sort of biological essence or any sort of "natural" distinction'; rather they are constructed and reproduced socially by a variety of actors, foremost of which is the state. Eric Hobsbawm (1992) reflects that people do not so much *remember* but are *reminded.* They are reminded of who they are by the selective prioritization of their heritage and identity, through various mediums – from the classroom to public space. Teaching history and geography, coding the built environment, celebrating national holidays, playing national anthems at sports events; national identity has been fostered in multiple ways, and geography has played a central role.

Territory, spatiality and geographical imagination have historically been essential elements in any nation or state's self-identification (Gruffudd, 1995; Graham, 1997, 2000). As James Martin (2005: 98) observes, spatiality is 'widely recognised as a key dimension in the formation of social identities', with national identity involving citizens' 'perception of the importance of territorial location and history in the formation of elements that make up their common identity'. To this end, representative landscape ideals often became metaphors for the nation and served to strengthen senses of national identity. For the Irish, such a representative landscape was embodied by the West of Ireland (Nash, 1993); for the English, the Cotswolds emerged as a prominent national landscape imaginary (Brace, 1999); and as John Agnew (1998) argues by using the example of Italy the absence of a national landscape ideal often coincided with the emergence of a weaker national identity. In other crucial ways, too, geography played a key role in the construction of national identity. A country's public space is a vital canvas through which to narrate and perform the nation. Cityscapes, townscapes and major historical sites became typically adorned with flags, monuments and statuary in the aftermath of independence, and newly designated national spaces became perennially (re)invested with meaning by political and cultural performances of memorialization (see also Section 5). Yvonne Whelan (2002), for instance, has shown how the iconography of Dublin's public space was recoded in the early 1920s after Ireland's War of Independence against Britain. Of course, in Ireland as elsewhere, long before independence, public space was utilized in the assertion of burgeoning national identity and defiance against colonial rule (see Chapter 2).

Geographically, any newly independent nation's capital city invariably became a central focus of national identity in the siting, building and symbolizing of various national institutions, such as museums and the national parliament (Atkinson and Cosgrove, 1998; Lorimer, 2002). For the United States of America, the national mall in its capital, Washington, emerged as the pivotal site of the nation. Stretching from the Capitol to the Lincoln Memorial, leading historical figures and past presidents are celebrated and the fallen in past wars are remembered (see Figure 5.1). Sharing space with centres of power such as the White House and the Capitol reinforces authority and agency, and collectively the various memorials and museums dotted along the mall are all key sites in the narration of American national identity. Any state, through its capital cityscape, possesses the power and forums to elevate narratives of national identity by investing meaning in purposeful politico-ideological and cultural (re)productions. However, it is important to bear in mind that in many regions throughout the world government attempts to culturally and ideologically order public space have been transcended by those that seek to disrupt the prioritized norm. Societies characterized by disparate historical, religious and cultural identifications and aggrievements, for example, frequently have a contested public space. This can be seen in the opposing wall murals of Belfast, Northern Ireland, or in the cityscape of Mostar, Herzegovina, where, in the aftermath of the Yugoslav Wars in the 1990s, a rebuilt and oversized Franciscan clock tower and giant Christian cross towering on the hilltop (Figure 5.2) have come to dominate the west bank of the Neretva river, while rebuilt Muslim minarets command the skyline on the other side.

Figure 5.1 Vietnam War Memorial, Washington DC, USA
(Photo: J. Morrissey, February 2006)

Figure 5.2 Franciscan clock tower and Christian cross, Mostar, Herzegovina
(Photo: J. Morrissey, August 2005)

Narrating the Nation

A *national meta-narrative*, re-telling a version of the past that connects people from across the nation, whether in triumph, loss or shared values, is vital to the construction of all national identities (Bhabha, 1990). Prioritized versions of national culture and identity are advanced in primary and secondary education, where senses of history are commonly essentialist (Morrissey, 2006); they are reproduced in the tourism and heritage industries by the celebration of specific artifacts and historical events (Kneafsey, 1998; Johnson, 1999); and they are reinforced in public space by representing and performing the most important elements of the past in the built environment (Hetherington, 1998; Atkinson et al., 1999). The state has the power to reify a selective historical meta-narrative in its (re)production of national identity and, to this end, key state mechanisms include national education, national institutions, national media and national iconography. All collectively contribute to what Michael Billig has outlined as the ubiquitous power of *banal nationalism* in our everyday lives: the 'continual "flagging", or reminding, of nationhood' (Billig, 1995: 8; see also Paasi, 1991). The raising of a national flag and exaltation of a past leader in public space, such as Chairman Mao in Tiananmen Square (Figure 5.3), are part of

Figure 5.3 Hoisting the national flag, Tiananmen Square, Beijing, China (Photo: J. Morrissey, April 2004)

Figure 5.4 Vojni Muzej, Belgrade, Serbia (Photo: J. Morrissey, August 2005)

what Billig means by banal nationalism. But there are of course other ways in which figures, events and stories from the past are connected to the present. Museums, for instance, typically elevate the most celebrated moments from a nation's past, while selecting out those that are inglorious or simply do not fit the nationalist self-image (Forty and Kuchier, 2001; Morrissey, 2005). In the *Vojni Muzej* (War Museum) in Belgrade (Figure 5.4), for example, the story of the 1990s is told as one of defiance against NATO, in which words like *genocide* or *Srebrenica* do not feature.

The construction of any national identity is underscored by selective memory and inaccuracy. As Ernst Renan famously declared at his Sorbonne lecture in Paris in 1882, '[f]orgetting, I would even go so far as to say historical error, is a crucial factor in the creation of a nation' (Renan, 1990: 11). The most cursory examination of any nation on earth would bear out Renan's contention (see, for instance, Morgan, 1984; Chapman, 1992; Harvey, 2003). Peter Taylor and Colin Flint (2007) use the case in point of Scotland and the Scottish kilt, one of the most widely known European national dresses, to highlight how senses of tradition and national identity can be effectively invented. They note that 'the tartan originates from the

Netherlands, that kilts come from England, and that there were no "clan tartans" before 1844'; the kilt, they argue, is 'a tradition that has been invented as part of a fabricated Scottish history' (Taylor and Flint, 2007: 165). Other elements of Scotland's retrospective invention of tradition have also been documented (see Trevor-Roper, 1984; Pittock, 1991; Broun et al., 1998).

Invented traditions have played an important role in the historical social construction of identity throughout the world (Hobsbawm, 1984). This has continued to the present. Benito Giordano (2000, 2001), for example, has shown how the separatist political party in contemporary northern Italy, *Lega Nord* (Northern League), has attempted to gain independence via the most fundamental methods of historical nation-building: fostering geographical imagination and forging banal nationalism. Since the mid-1990s, Lega Nord has invented a region they call *Padania* (see Figure 5.5) and have posited

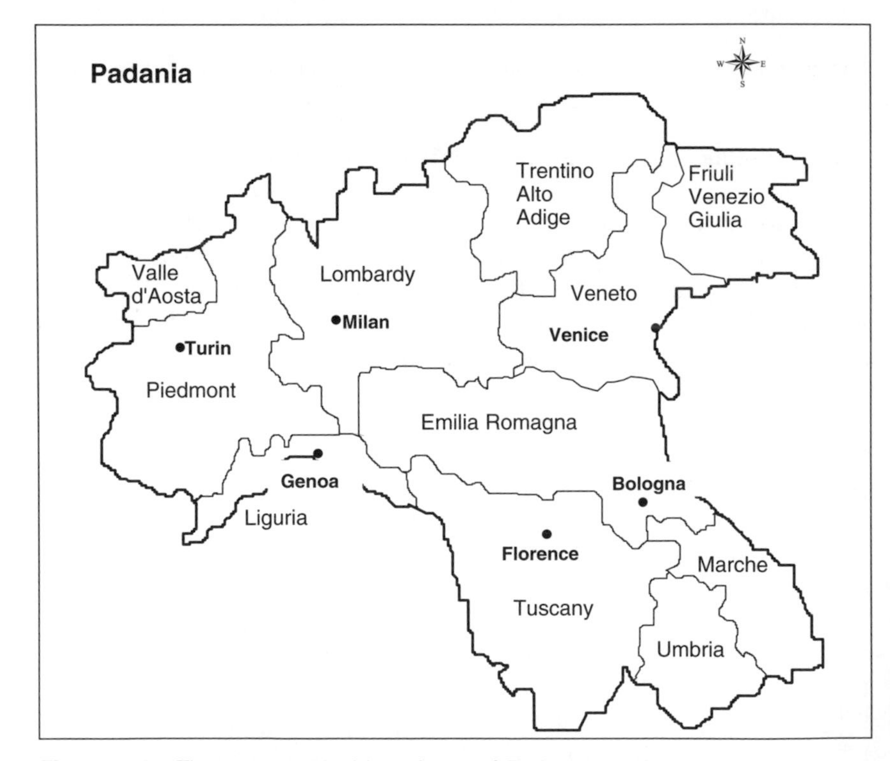

Figure 5.5 The geographical imaginary of Padania, northern Italy

it as an historic geographic area with an ancient Celtic ethos. The idea of Padania has been used to reclaim this alleged history and geography and, in so doing, to legitimize and lend an air of antiquity to the territorial and political claims of Lega Nord (Cento Bull and Gilbert, 2001). To this end, various techniques of banal nationalism have been used at different sites of representation and performance to normalize the idea of the new geographical imaginary – from flags, symbols, associations and festivals to newspapers, radio, television and the internet. Lega Nord and their construct of Padania provide us with an excellent example of the manner in which senses of identity (in this case, aspiring national identity) can be historically identified and imagined and then reinforced in banal social and cultural geographies. As Giordano (2001: 36) notes, the 'invention of "Padania" shows that such identities do not necessarily have to be inherited from primordial roots but can actually be created almost from scratch in contemporary society'.

The notion of invented traditions ties in closely with Benedict Anderson's idea of *imagined communities* (Anderson, 1991). In his reflection on the origins and spread of nationalism, he argues that most people in any given nation will never meet the vast majority of their fellow citizens, yet still maintain an allegiance to them. For Anderson, senses of national, communal identity are largely *imagined*. Although he does acknowledge the import of print cultures such as maps (which are, of course, material and real), he has been charged with placing 'too much emphasis on the idea of the nation rather than on the historical geographies of nationhood' (Katz, 2003: 251). In other words, his thesis is problematic in that he does not allow for the tangible *geographies* of nationalism that (re)produce communal identity in *real* ways (for further discussion on the geographical (re)productions of nationalism, see Jones and Fowler, 2007). Political, social and cultural (re)productions of the state may be selective and in some cases erroneous, as outlined above, but their spatial and material consequences are no less effective in the inculcation and maintenance of national identity. It is important to remember, too, that the state, as a typically centralized apparatus, commands the power and resources to build senses of unity around fundamental administrative and institutional frameworks that have often been in place since colonial and earlier periods.

National identity is also advanced in very real ways outside of the sphere of the state, via collective religious worship, for example, or the communal networking of music, song and oral history. In the end, of course, it is real people and their traditions that enact senses of national identity. In the former Soviet state of Belarus, the first thing that newly-weds do immediately after the wedding ceremony is to go with their families and friends to the nearby World War II memorial to honour fallen loved ones from the local community. Millions of Belarusians died repelling the German advance on Moscow in World War II and this unprecedented national loss is remembered from the capital, Minsk, to little villages across the nation (see Figure 5.6). These now sacred geographical spaces are active nodes in maintaining and networking senses of communal national identity that foster strong ties to the land and the fallen of the past that defended it.

Figure 5.6 World War II memorial, Goradiche, southern Belarus (Photo: J. Morrissey, July 2002)

Remembering communal loss, trauma and tragedy is a key element in the maintenance of national identities. The millions of Irish-Americans who have marched on St Patrick's Day every year for more than two centuries – in towns and cities across America, from Boston to San Francisco, from New Orleans to Chicago – have been variously connecting to an Irish *emigrant* experience that is commonly envisioned via a tragic Irish history of famine and loss. This meta-narrative (like all others) is, of course, partly imagined and essentialist, and, in losing sight of the myriad intricate human geographies and histories of the emigrants themselves, it also frequently reduces them to victims. In the similar Scottish-American context, Paul Basu (2007) notes how the dominant, loss-centred memory or *victim narrative* of the Great Highland Clearances serves to disallow the complexities of the Scottish Highland diasporic experience and to deny the agency of the emigrants in their own lives.

Notwithstanding these important points, our pasts tend to become mythologized as stories that work, that connect people, that support. And loss matters. The Great Irish Famine of the late 1840s, which saw over a million people die of starvation and even more emigrate to the United States and elsewhere, is remembered in numerous ways in Irish America, from pub songs and popular culture to public monuments. The opening of the Irish Hunger Memorial in downtown Manhattan (see Figure 5.7) in 2002 by New York Governor George Pataki and President of Ireland Mary McAleese demonstrates the continued importance of historical geographies in the present in (re)producing and networking senses of communal identity (in this case, diasporic Irish identity). The memorial, which is situated on a quarter-acre of prime real estate, adjacent to the World Trade Center site, features a carefully relocated famine cottage from County Mayo in north-western Ireland (symbolically with soil, flora and rocks from each of the other 31 counties of Ireland). As you enter the site, the aural narration of aspects of Ireland's *Gorta Mór* (Great Hunger) in the nineteenth century connects too in solidarity with all regions in the contemporary world experiencing the horrors of famine. Finally, of course, for other nations, too, remembering loss is a crucial element of national identity. For Israelis and Jews throughout the world, for example, the former Nazi concentration camps of World War II, such as Auschwitz in southern Poland (Figure 5.8), are sacred spaces in which to communally pray for the millions murdered in the Holocaust and to remember the terror of biopolitical control.

Figure 5.7 Irish Hunger Memorial, New York City, USA (Photo: J. Morrissey, October 2007)

Figure 5.8 Auschwitz II-Birkenau Concentration Camp, Oswiecim, Poland (Photo: J. Morrissey, August 2005, published courtesy of Auschwitz Memorial)

Sameness, Difference and the Complexities of Identity Networks

A key requirement of all national identities is the dual identification of sameness and difference, or *Self* and *Other*. As Cindi Katz (2003: 249) observes, identities are 'relational', with the Self 'always defined in terms of what it is not', the Other. In early modern Protestant England, for example, the emerging English nationalism was defined by a fearful anti-Catholicism against what it was not – Catholic Irish, Scottish or French (Marotti, 1997; Colley, 2005). Otherness was historically produced and disseminated in a variety of ways in the colonial past, as outlined in the opening two chapters in discussions around the work of Edward Said and Homi Bhabha. Said's critique of colonial discourse and cultural ascription and Bhabha's writing on in between space and hybridity can also be applied, however, in the related context of nationalism, where the historical discursive championing of specific tropes of national identity simultaneously obscured cultural complexities and marginalized senses of difference (Said, 1978; Bhabha, 1994).

National identities were typically produced through a variety of asymmetrical relations; their interrogation, therefore, necessarily involves consideration of the exclusionary practices, discrimination and violence in the social spaces of the past (Maalouf, 2001; Sen, 2006). Historical geographers have explored the social exclusion and conflict inherent in senses of national identity in multiple political, economic and cultural contexts (see Section 3). In examining historical geographies of racism, for instance, geographers have been particularly concerned with the spatial mechanisms of political rule through which social power and cultural hegemony were underpinned. In exploring issues of resistance, they have also highlighted the counter-hegemonic strategies and practices in which 'uneven power relations' were 'interrupted, compromised and undone on spatial as well as social and political grounds' (Katz, 2003: 260; see also Morrissey, 2003).

Despite the power of the state, multiple identity networks on various scales have historically disrupted prioritized senses of national identity across the world. Since the unification of Italy, for example, in the late nineteenth century, the Italian state's attempts to construct Italian national identity have taken place in the shadow of much older and stronger regional and local identifications throughout the country. Understanding Italian national identity and its historical development is impossible without consideration of the complicated intersections of other scalar identities. Linear models of national identity also negate other cultural complexities within historical societies. Different identity types – based on gender, age, sexuality, class and so on – co-existed, networked and transmuted each other in multi-faceted ways, both within and beyond the nation, and historical geographers have been particularly concerned with sifting out these intricacies and narrating more nuanced accounts of the past (Blunt and Rose, 1994; Mulligan, 2002). To this end, the notion of *hybridity* has been especially useful in contesting simplified models of national identity. It has served, for example, as 'a potent concept of resistance' to 'essentialising narratives of nation and race', and as a 'trenchant critique of modernist binaries and normative assumptions based on age-old notions of separation and linearity' (Mitchell, 2005: 192).

Conclusion

Geography has played a central role in the historical construction, performance and reproduction of national identity throughout the world. Using different examples, this chapter has sought to outline the role of geographical imagination, territoriality, spatiality and public space in the historical process of nation-building. It has been concerned too with problematizing reductive nationalist representations of historical societies that mask key nuances from the geographical worlds of the past, including cultural complexity, hybridity and co-existing scalar identities. In stressing how essentialist models of national identity promulgated Otherness, attention has also been drawn to the spatial mechanisms that facilitated exclusionary social practices and political and cultural hegemony. Finally, in underlining the historical social construction of all national identities, emphasis has been placed on the selectivity of prioritized national meta-narratives of heritage and culture. Recognizing the historical relativism of all forms of national identity is one of the most effective ways of challenging the emergence in the modern world of absolute senses of identity whose endgame is so frequently racism, discrimination and conflict.

KEY POINTS

- Geographical imagination, territoriality and spatiality have been essential elements in the historical construction of national identity.
- In the process of nation-building, a nation's public space is a vital canvas through which to narrate and perform prioritized and selective meta-narratives of identity.
- Historical political and socio-cultural constructs of the nation may be selective and in some cases erroneous, but tangible geographies of nationalism have (re)produced communal identity in real ways, in which communal trauma and loss have been key elements.
- The historical discursive championing of essentialist tropes of national identity has simultaneously promulgated Otherness and elite cultural hegemony.
- National identities have been typically produced through a variety of asymmetrical social relations and their interrogation necessarily involves historical examination of exclusionary practices, discrimination and violence.
- Nationalist representations of historical societies mask key nuances from the geographical worlds of the past, including cultural complexity, hybridity and different scalar identities co-existing both within and beyond the nation.

References

Agnew, J. (1998) 'European landscape and identity', in B. Graham (ed.) *Modern Europe: Place, Culture and Identity*. London: Arnold, pp. 213–235.

Anderson, B. (1991) *Imagined Communities: Reflections on the Origins and Spread of Nationalism*. London: Verso.

Atkinson, D. and Cosgrove, D. (1998) 'Urban rhetoric and embodied identities: city, nation and empire at the Vittorio Emanuele II monument in Rome, 1870–1945', *Annals of the Association of American Geographers*, 88(1): 28–49.

Atkinson, D., Cosgrove, D. and Notaro, A. (1999) 'Empire in modern Rome: shaping and remembering an imperial city', in F. Driver and D. Gilbert (eds) *Imperial Cities: Landscape, Display and Identity*. Manchester: Manchester University Press, pp. 40–63.

Basu, P. (2007) *Highland Homecomings: Genealogy and Heritage Tourism in the Scottish Diaspora*. London: Routledge.

Bhabha, H. (1990) 'Introduction: narrating the nation', in H. Bhabha (ed.) *Nation and Narration*. London: Routledge, pp. 1–7.

Bhabha, H. (1994) *The Location of Culture*. London: Routledge.

Billig, M. (1995) *Banal Nationalism*. London: Sage.

Blunt, A. and Rose, G. (eds) (1994) *Writing Women and Space: Colonial and Postcolonial Geographies*. New York: Guilford.

Brace, C. (1999) 'Finding England everywhere: regional identity and the construction of national identity, 1890–1940', *Ecumene*, 6(1): 90–109.

Broun, D., Finlay, R.J. and Lynch, M. (eds) (1998) *Image and Identity: The Making and Re-Making of Scotland Through the Ages*. Edinburgh: J. Donald.

Castells, M. (1997) *The Power of Identity*. Oxford: Blackwell.

Cento Bull, A. and Gilbert, M. (2001) *The Lega Nord and the Northern Question in Italian Politics*. New York: Palgrave.

Chapman, M. (1992) *The Celts: The Construction of a Myth*. Basingstoke: Macmillan.

Colley, L. (2005) *Britons: Forging the Nation, 1707–1837* (2nd edn). New Haven, CT: Yale University Press.

Forty, A. and Kuchier, S. (eds) (2001) *The Art of Forgetting*. Oxford: Berg.

Giordano, B. (2000) 'Italian regionalism or "Padanian" nationalism – the political project of the Lega Nord in Italian politics', *Political Geography*, 19(4): 445–471.

Giordano, B. (2001) 'The contrasting geographies of "Padania": the case of the Lega Nord in northern Italy', *Area*, 33(1): 27–37.

Graham, B. (1997) 'The imagining of place: representation and identity in contemporary Ireland', in B. Graham (ed.) *In Search of Ireland: A Cultural Geography*. London: Routledge, pp. 192–212.

Graham, B. (2000) 'The past in place: historical geographies of identity', in B. Graham and C. Nash (eds) *Modern Historical Geographies*. Harlow: Longman, pp. 70–99.

Gruffudd, P. (1995) 'Remaking Wales: nation-building and the geographical imagination', *Political Geography*, 14(3): 219–239.

Harvey, D.C. (2003) '"National" identities and the politics of ancient heritage: continuity and change at ancient monuments in Britain and Ireland, c. 1675–1850', *Transactions of the Institute of British Geographers*, 28(4): 473–487.

Hastings, A. (1997) *The Construction of Nationhood: Ethnicity, Religion and Nationalism*. Cambridge: Cambridge University Press.

Hetherington, K. (1998) *Expressions of Identity: Space, Performance, Politics*. London: Sage.

Hobsbawm, E. (1984) 'Introduction: inventing traditions', in E. Hobsbawm and T. Ranger (eds) *The Invention of Tradition*. Cambridge: Cambridge University Press, pp. 1–14.

Hobsbawm, E. (1992) *Nations and Nationalism Since 1780: Programme, Myth, Reality*. Cambridge: Cambridge University Press.

Johnson, N. (1999) 'Framing the past: time, space and the politics of heritage tourism in Ireland', *Political Geography*, 18(2): 187–207.

Jones, R. and Fowler, C. (2007) 'Placing and scaling the nation', *Environment and Planning D: Society and Space*, 25: 332–354.

Katz, C. (2003) 'Social formations: thinking about society, identity, power and resistance', in S.L. Holloway, S.P. Price and G. Valentine (eds) *Key Concepts in Geography*. London: Sage, pp. 249–265.

Kneafsey, M. (1998) 'Tourism and place identity: a case-study in rural Ireland', *Irish Geography*, 31(2): 111–123.

Lorimer, H. (2002) 'Sites of authenticity: Scotland's new parliament and official representations of the nation', in D.C. Harvey, R. Jones, N. McInroy and C. Milligan (eds) *Celtic Geographies: Old Cultures, New Times*. London: Routledge, pp. 91–109.

Maalouf, A. (2001) *In the Name of Identity: Violence and the Need to Belong*. New York: Arcade.

Marotti, A.F. (1997) 'Southwell's remains: Catholicism and anti-Catholicism in Early Modern England', in C.C. Brown and A.F. Marotti (eds) *Texts and Cultural Change in Early Modern England*. Basingstoke: Macmillan, pp. 37–65.

Martin, J. (2005) 'Identity', in D. Atkinson, P. Jackson, D. Sibley and N. Washbourne (eds) *Cultural Geography: A Critical Dictionary of Concepts*. London: I.B.Tauris, pp. 97–102.

Mitchell, K. (2005) 'Hybridity', in D. Atkinson, P. Jackson, D. Sibley and N. Washbourne (eds) *Cultural Geography: A Critical Dictionary of Concepts*. London: I.B.Tauris, pp. 188–193.

Morgan, P. (1984) 'From a death to a view: the hunt for the Welsh past in the Romantic period', in E. Hobsbawm and T. Ranger (eds) *The Invention of Tradition*. Cambridge: Cambridge University Press, pp. 43–100.

Morrissey, J. (2003) *Negotiating Colonialism*. London: HGRG, Royal Geographical Society.

Morrissey, J. (2005) 'A lost heritage: the Connaught Rangers and multivocal Irishness', in M. McCarthy (ed.) *Ireland's Heritages: Critical Perspectives on Memory and Identity*. Aldershot: Ashgate, pp. 71–87.

Morrissey, J. (2006) 'Ireland's Great War: representation, public space and the place of dissonant heritages', *Journal of Galway Archaeological and Historical Society*, 58: 98–113.

Mulligan, A. (2002) 'A forgotten "Greater Ireland": the transatlantic development of Irish nationalism', *Scottish Geographical Journal*, 118(3): 219–234.

Nash, C. (1993) '"Embodying the nation": the West of Ireland landscape and Irish identity', in B. O'Connor and M. Cronin (eds) *Tourism in Ireland: A Critical Analysis*. Cork: Cork University Press, pp. 86–112.

Paasi, A. (1991) 'Deconstructing regions: notes on the scales of spatial life', *Environment and Planning A*, 23: 239–254.

Pittock, M. (1991) *The Invention of Scotland*. London: Routledge.

Renan, E. (1990) 'What is a nation?', in H. Bhabha (ed.) *Nation and Narration*. London: Routledge, pp. 8–22.

Said, E. (1978) *Orientalism: Western Conceptions of the Orient*. New York: Pantheon.

Sen, A. (2006) *Identity and Violence: The Illusion of Destiny*. New York: W.W. Norton.

Taylor, P. and Flint, C. (2007) *Political Geography: World-Economy, Nation-State and Locality* (5th edn). Harlow: Pearson.

Trevor-Roper, H. (1984) 'The invention of tradition: the Highland tradition of Scotland', in E. Hobsbawm and T. Ranger (eds) *The Invention of Tradition*. Cambridge: Cambridge University Press, pp. 15–41.

Whelan, Y. (2002) 'The construction and destruction of a colonial landscape: monuments to British monarchs in Dublin before and after independence', *Journal of Historical Geography*, 28(4): 508–533.

FURTHER READING

Anderson, B. (1991) *Imagined Communities: Reflections on the Origins and Spread of Nationalism*. London: Verso.

Bhabha, H. (1994) *The Location of Culture*. London: Routledge.

Graham, B. (1997) 'The imagining of place: representation and identity in contemporary Ireland', in B. Graham (ed.) *In Search of Ireland: A Cultural Geography*. London: Routledge, pp. 192–212.

Katz, C. (2003) 'Social formations: thinking about society, identity, power and resistance', in S.L. Holloway, S.P. Price and G. Valentine (eds) *Key Concepts in Geography*. London: Sage, pp. 249–265.

6 IMAGINATIVE GEOGRAPHIES AND GEOPOLITICS

John Morrissey

Introduction

Soon after the September 11 attacks on New York and Washington in 2001, the Italian philosopher Giorgio Agamben wrote that it is 'the task of democratic politics to prevent the development of conditions which lead to hatred, terror and destruction and not to limit itself to attempts to control them once they have already occurred' (2001). Agamben points here to the general failure of the Western democratic system to prevent the conditions of ignorance, injustice and inequality that lead to conflict and war. Many of the most critical opponents of the current war on terrorism launched in response to the September 11 attacks point to its selective mapping of a geography of danger that serves to bury historical injustices, prior western interference and contemporary geoeconomic interests under a prevailing vernacular of *terror* and *threat*. In a world where *geographical representation* is key, historical geographers can play a critical role in historicizing and geo-graphing places and events, and in so doing dismantle reductive scriptings of distant conflicts. This chapter uses the example of the *war on terror* to illuminate the role of *affective imaginative geographies* in the legitimizing and waging of war. It initially outlines the historical context of the use of imaginative geographies in the envisioning and representation of conflict before reflecting on the various means by which geographers have critically challenged contemporary abstracted scriptings of war and geopolitics.

Imaginative Geographies and Affect

Throughout human history, geographical imaginings of distant and different *Others* have been employed to both rationalize and frame the waging of war. From the Persian Wars to the Crusades, from the Wars of the Three Kingdoms to the Yugoslav Wars, conflict has been imagined and enacted via purposeful rudimentary mappings of territory and identity, civility and barbarism, threat and terror (war is also typically remembered via simplified representation and performance; see Section 5). Such discourses frequently submerge underpinning geoeconomic and geopolitical imperatives, and instead utilize *imaginative geographies* to present an essentialist rationale for the use of force. Geo-graphing and prioritized geographical knowledges, in other words, perform central roles in the practice of war.

Imaginative geographies are produced and disseminated at multiple sites/forums of representation and performance: from newspapers to TV news; from radio to film; from press conferences to public marches; from state documents to the internet (for discussion on these and other sources, see Chapter 24). Historically, the circulation of imaginative geographies has served to collectively legitimate a vocabulary of difference in which specific tropes of identity in the form of 'representative figures' have come to stand for whole nations (Said, 2003 [1978]: 71). Stretching back centuries in Europe, for example, the representative figure of Islam symbolized 'terror, devastation, the demonic' (Said, 2003 [1978]: 59): 'For Europe, Islam was a lasting trauma. Until the end of the seventeenth century the "Ottoman peril" lurked alongside Europe to represent for the whole of Christian civilization a constant danger'. Elsewhere, equally representative imaginative geographies have established a powerful register of difference, ignorance, fear and hate. From Ancient Rome to the new worlds of the Americas, the barbarians, the uncivilized, the perennial threat lurked just *beyond the pale*.

In the modern world, mainstream public commentators on the September 11 attacks have commonly sought to conceptualize them via a discourse of Islamic *terror* (Silberstein, 2002; Chomsky, 2003; Weber, 2003; for more on visual cultures, see Chapter 23). Outside of the sufferings and heroics of the victims and their families, the focus of Western attention has typically been on the identification and punishment of the individuals and more importantly *types* responsible – those who *hate us* and *our values*. This prevailing reductive formula for understanding the so-called *post-9/11* world is based on *affective* imaginative geographies that justify and indeed demand retaliatory action in the form of aggression. As Gearóid Ó Tuathail argues, the abiding consequence of the essentialist American reaction to 9/11 was that the decision to invade Iraq was based more on *affect* than intellect:

> The triumph of affect over intellect is marked by the desire to attack Iraq even though there is not convincing evidence for doing so. Intellectual deliberation and policy credibility take a back seat to "instinctive" convictions and prejudgements. Saddam Hussein is an "evil-doer," and in the scaled down world of affect, this alone justifies "regime change." (2003: 853)

In the oft-cited *clash of civilizations* between *East* and *West* today, each side typically refers to the other as *perpetrator* and *evil*, while identifying themselves as *victim* and *good* (the designations *East* and *West* are, of course, reductive and hugely problematic; nonetheless, they are commonly employed as givens – see, for example, Huntingdon, 1993). It is important to remember that affective imaginative geographies not only function in the so-called West; a mythical picture of a corrupted, infidel and morally doomed West is a common touchstone of essentialist populist discourses in the so-called East. Clearly, however, this contemporary polemical language war masks a longer history of Western colonialism and strategic interventions in the Middle East, and an accompanying sequence of violence and reprisal. Neil Smith (2003: xi), for example, reminds us that 'the historical geography of American globalism has everything to do with the first major war of the twenty-first century, the so-called war on terrorism'. America's pragmatic political, economic and military support for regimes favourable to US interests in the Middle East (since at least World War II), the long-standing Israeli–Palestinian conflict (in which the US is seen in the Arab world to overwhelmingly support Israel) and the recent proliferation of US bases on Arab soil throughout the region are all connected to both the 9/11 attacks and ongoing resistance to the US-led war on terror. In other words, when assessing and critiquing the political geography and violence of the contemporary Middle East, recognizing and contextualizing its historical geographies is vitally important.

In prompting us 'to rethink the lazy separations between past, present, and future', Derek Gregory has been foremost in critiquing the powerful historical and contemporary imaginative geographies that continue to shape our world (2004: 7). Gregory's *The Colonial Present* illuminates recent Western interventions overseas as underpinned by an enduring register of essentialist imaginative geographies: West versus East; good versus evil; civility versus barbarity (2004). Using the examples of the US-led invasions of Iraq and Afghanistan and Israel's war in Palestine, Gregory demonstrates how specific geographical representations or *geopolitical projections* serve to not only discursively separate *us* from *them*, but in real terms to dictate the logics of Western military intervention and to allow for the subjects of Western violence to be rendered as mere objects; whose corporeality has no resonance in the west. As Naomi Klein argues, Iraqi deaths (like those of Afghanis and Palestinians) simply 'don't count' (2004). The extraordinary rendition of tens of thousands to the status of *bare life* in camps in Iraq, Afghanistan, Palestine, Guantanamo Bay and elsewhere reveal the extent to which the omnipresent discursive spacings between *us* and *them* in everyday life have allowed for, and indeed legitimized, the suspension of the most basic human rights of citizens of sovereign states throughout the world (Agamben, 1998, 2005).

CENTCOM's War on Terror: Geopolitical and Geoeconomic Logics

The following example of United States Central Command (CENTCOM) demonstrates further the links between geopolitical projection and geopolitical practice. Since its initiation in 1983, CENTCOM has been responsible for the military planning, coordination and implementation of US grand strategy in the Middle East. Seen in Figure 6.1, CENTCOM's 'Area of Responsibility' (AOR) encompasses the world's most energy-rich region, including Iran, Iraq, Kuwait and Saudi Arabia (on 1 October 2008, all of the command's African countries, with the exception of Egypt, became officially part of a new Africa Command). Given its geographic focus, CENTCOM has emerged in recent years as the chief body responsible for the military operation of the US-led war on terror. Accordingly, it has produced various strategic geopolitical projections that have centrally informed the formulation of policy. Its 'theater strategy' document *Shaping the Central Region for the 21st Century*, for example, outlines its key objective: 'Protect, promote and preserve U.S. interests in the Central Region to include the free flow of energy resources, access to regional states, freedom of navigation, and maintenance of regional stability' (US CENTCOM).

CENTCOM's historical geopolitical scripting of the Middle East sheds important light upon the strategic and geoeconomic priorities that underlie the so-called war on terror. In addition to the geographic focus on the Middle East, the extension of its AOR in 1999 to include the oil-, gas- and mineral-rich Central Asian states of Kazakhstan, Kyrgyzstan, Tajikistan, Turkmenistan and Uzbekistan only makes sense if a key purpose of CENTCOM is a specialization on the geopolitics of energy. The importance of the Central Asian states is clear from CENTCOM Commander General John P. Abizaid's Statement before the Senate Arms Services Committee in 2006: 'In a region at the crossroads between Europe and Asia, the stability and further development of transportation and energy networks is increasingly important for global economic health' (US Senate Armed Service Committee, 2006: 41–42).

For the war in Iraq, stated motives about weapons of mass destruction dominated the media networks and public opinion. However, other

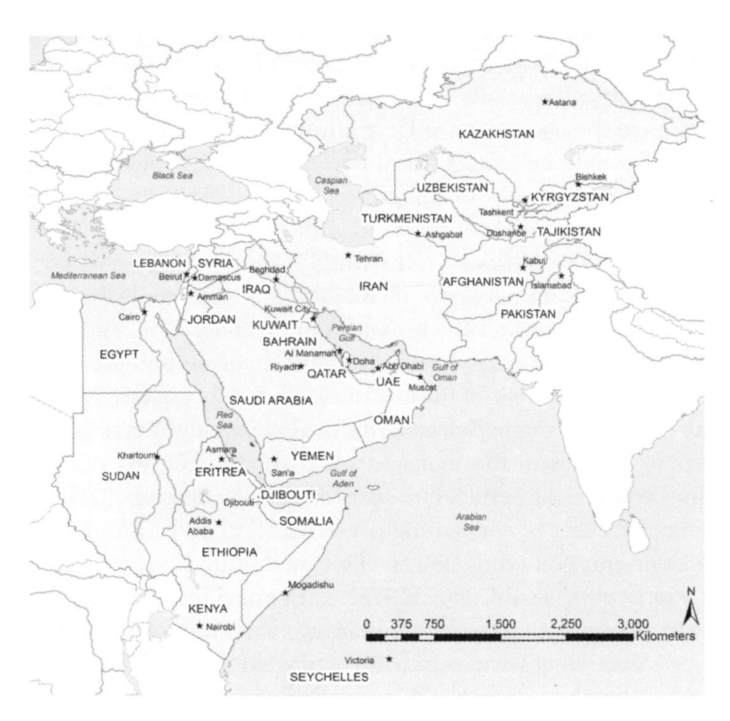

Figure 6.1 CENTCOM Area of Responsibility, July 2008 (Adapted from US Department of Defense, 2007)

motives not stated about business opportunities were certainly also present (for more on this, see Chapter 17 of Naomi Klein's *The Shock Doctrine*). The key point here is that military knowledge is commonly not challenged with respect to what it conceals in terms of policy. CENTCOM's scripted strategy functions in a world in which military '"policy" is consistently (and increasingly) used to override and even suppress debate' (Gregory, 2005: 183; see also Harvey, 2004). Military representations are rarely contested in the public realm, yet their interrogation and illumination are crucial to understanding associated geopolitical and geoeconomic practices (Johnson, 2004). If we look closer at CENTCOM's war on terrorism, we can see that it is based on a loosely scripted 'cartography of danger' that has emerged from a geopolitical abstraction whose reductive formula has rendered 'peoples and places ready for military action' (Dalby, 2007; see also Graham, 2005). This contextualization and representation of the Middle East serves, of course, to deliberately negate the humanity and complexity of the largely unseen human geographies on the ground. CENTCOM's war and actions, however, are also underpinned and allowed for by the careful designation of its AOR as a strategic space in the world vital to *global economic health* and the success of neoliberalism (see Gold, 1988).

The geographical representations of the Middle East by CENTCOM not only legitimize the operations of US military grand strategy but also form part of a wider political and cultural discourse on the region that has produced a moral cartography or imaginary constituted of distinct structures of identity, difference and terror (Said, 1997; Shapiro, 1997). The ongoing presence of CENTCOM forces in the Middle East – replete with one of the most extensive basing networks in history – is commonly legitimated in US government circles and the broader public sphere via a simplified discourse of risk and insecurity that centrally involves imaginative geographies. Edward Said, in one of his last works before his death, warned of the enduring power of imaginative geographical knowledges in framing understandings of conflict in the modern world. In the context of the United States in the post-9/11 period, he observed the 'hardening of attitudes, the tightening of the grip of demeaning generalization and triumphalist cliché' and the 'dominance of crude power allied with simplistic contempt of dissenters and "others"' (Said, 2003 [1978]: xiii). For Said, projections of grand strategy from imperialism to contemporary geopolitics were made possible by 'the construction of various kinds of knowledge, all of them in one way or another dependent upon the perceived character and destiny of a particular geography' (1993: 93).

Military Geography, Critical Geopolitics and Counter Geographies

Over the last decade, in particular, the traditional sub-discipline of military geography has been buoyed by advances in geographic information system (GIS) technology that have attracted lucrative research and development opportunities in the areas of defence and surveillance. As a rejoinder to this largely uncritical *strategic studies* research, many geographers have engaged military geographies in an oppositional and critical manner (Woodward, 2004; Farish, 2007). For the first Gulf War, for example, James Sidaway (1998) has outlined the quickly mobilized Orientalist representational strategies of the West that were used to reduce a previously heralded secular and progressive Iraq to a place of tyranny personified by Saddam Hussein. In more recent years, geographers have offered a range of critical geopolitical accounts of the global war and terror (Agnew, 2003; Harvey, 2003; Jhaveri, 2004; Dodds, 2005; Smith, 2005; Dalby, 2006; David and Grondin, 2006; Hyndman, 2007; Fluri, 2009; Morrissey, 2009, 2011a). One of the key elements of much of this work, in what can be loosely described as *critical geopolitics*, is the illumination of the import of geographical knowledges (particularly Manichean imaginative geographies of good and evil) for the effective operation of hegemonic power (for related work outside of geography, see: Atkinson, 1999; Dodge, 2003). Critical geopolitics has sought to disrupt and interrogate the basis, legitimization and operation of contemporary Western geopolitical power and knowledge (Ó Tuathail, 1996). A key concern has been to proffer more humane and nuanced *counter-geographies* that insist on the spatiality and materiality of global space. For the first Gulf War, for example, Gearóid Ó Tuathail has divulged the effacement of the material geographies – the corporeal realities beneath the cluster bombs and so-called smart bombs – by the bloodless *electronic spatiality* of the military machine and a complicit and submissive Western media and audience (1993). For the war on terror, Derek Gregory has shown the necessity of 'a cultural mobilization' that employs 'imaginative geographies' as 'powerful rhetorical weapons' folding 'difference into distance' (2005: 186). Gregory argues passionately for the scripting of counter-geographies that insist on real places with real people, with rights and bodies just like us.

Some of the most important counter-geographies concerning the war on terror have been advanced outside the academy, such as those seen in the unprecedented global public protests against the war in Iraq in 2003. Counter-geographies highlighting key issues such as the geoeconomic strategic importance of Iraq, the longer history and broader geography of Western interventionism in the region and the cultural complexities and human suffering effaced from sanitized politico-ideological and military rationales for the war were forwarded throughout the world. In this context, many geographers played important roles as public intellectuals in circulating critiques of the war (and broader war on terror) in a variety of public forums, from critical newspaper and internet articles to solidarity campaigns and public lectures. The dissemination of critical voices of reason, empathy and protest may not have stopped the war in Iraq but they were important to efforts to bring it to an end. The thousands of *pace* (peace) flags (see Figure 6.2) flying the length and breadth of Italy in the lead-up to the war did not initially prevent Italian troops being deployed but their continued presence in public space as the war raged on mirrored the country's huge opposition to the war, and the persistent calls for troop withdrawal were eventually acted upon after a change of government in 2006.

Figure 6.2 *Pace* protest, Cagliari, Italy, May 2003 (Photo: J. Morrissey)

Within the academy, recent years have witnessed a sustained mobilization of research in historical, political and cultural geography that has critiqued the operation of Western overseas power and the exercise of force in at least four ways: (i) first, geographers have sought to reveal the historical geographical contexts of contemporary geopolitics and geoeconomics (Farish, 2003; Jhaveri, 2004; Kearns, 2006); (ii) second, attention has been paid to rendering visible the violent geographies of Western military interventions (Flint, 2005; Gregory and Pred, 2007; Hyndman, 2007); (iii) third, focus has been placed on the predominant geographical discourses of us and them justifying and buttressing the advancement of Western military interventions (Coleman, 2004; Bialasiewicz et al., 2007; Dalby, 2007); and (iv) finally, informed by the work of Giorgio Agamben on bare life and the state of exception and Michel Foucault's writings on biopower and governmentality, geographers have illuminated the modalities of Western biopolitics in the so-called global war on terror (Foucault, 2007, 2008; Gregory, 2007; Morrissey, 2011b).

Conclusion

Imaginative geographies function in various ways in modern society in identifying and delineating difference. Maintained by fundamental relations of power and knowledge, they frequently serve to disconnect peoples and maintain ignorance, fear and hatred. The initiation of the war on terror demanded the mobilization and maintenance of simplified, accessible, mythical knowledges – imaginative geographies that compelled us to neatly take sides and confidently answer such questions as *are you with us or against us* (Hedges, 2003). The affective imaginative geographies of the war on terror continue to take our attention and empathy away from a whole series of historical and contemporary human geographies that need to be rendered visible. A more nuanced and humane orientation of critical geographical knowledges can help to counter simplified and politicized scriptings of both foreign conflict and foreign policy. To this end, historical geographers can play an important role in narrating the unseen contexts of contemporary geopolitics and revealing the framed imaginative geographies that have such potent power to render us overwhelmed – through words, images and affect if not through bombs.

KEY POINTS

- In a world where *geographical representation* is key, historical geographers can play a critical role in historicizing and geo-graphing places and events, and in so doing dismantle reductive scriptings of distant conflicts.
- Historically, the circulation of *imaginative geographies* has served to collectively legitimate a vocabulary of difference in which specific tropes of identity have come to stand for whole nations.
- Hegemonic geographical representations serve to not only discursively separate *us* and *them* but in real terms to legitimize and dictate the logics of military intervention and the exercise of violence.
- Official US military and geopolitical scriptings of the Middle East such as that of CENTCOM shed important light upon the *strategic and geoeconomic priorities* that underlie the war on terror.
- Critical geopolitics has sought to disrupt and interrogate the basis, legitimization and operation of contemporary Western geopolitical power and knowledge, and a key concern has been to proffer more humane and nuanced *counter-geographies* that insist on the spatiality and materiality of global space.
- The *affective* imaginative geographies of the current *war on terror* continue to take our attention and empathy away from a whole series of historical and contemporary human geographies that need to be rendered visible.

References

Agamben, G. (1998) *Homo Sacer: Sovereign Power and Bare Life* (trans. D. Heller-Roazen). Stanford, CA: Stanford University Press.

Agamben, G. (2001) 'On security and terror', *Frankfurter Allgemeine Zeitung*, 20 September.

Agamben, G. (2005) *State of Exception* (trans. K. Attell). Chicago: University of Chicago Press.

Agnew, J. (2003) 'American hegemony into American empire? Lessons from the invasion of Iraq', *Antipode*, 35(5): 871–885.

Atkinson, P. (1999) 'Representations of conflict in the western media: the manufacture of a barbaric periphery', in T. Skelton and T. Allen (eds) *Culture and Global Change*. London: Routledge, pp. 102–108.

Bialasiewicz, L., Campbell, D., Elden, S., Graham, S., Jeffrey, A. and Williams, A.J. (2007) 'Performing security: the imaginative geographies of current US strategy', *Political Geography*, 26(4): 405–422.

Chomsky, N. (2003) *Pirates and Emperors, Old and New: International Terrorism in the Real World.* Cambridge, MA: South End Press.

Coleman, M. (2004) 'The naming of terrorism and evil outlaws: geopolitical place-making after 11 September', in S. Brunn (ed.) *11 September and Its Aftermath: The Geopolitics of Terror.* London: Frank Cass, pp. 87–104.

Dalby, S. (2006) 'Geopolitics, grand strategy and the Bush doctrine: the strategic dimensions of U.S. hegemony under George W. Bush', in C.-P. David and D. Grondin (eds) *Hegemony or Empire? The Redefinition of American Power under George W. Bush.* Aldershot: Ashgate, pp. 33–49.

Dalby, S. (2007) 'The Pentagon's new imperial cartography: tabloid realism and the war on terror', in D. Gregory and A. Pred (eds) *Violent Geographies: Fear, Terror and Political Violence.* New York: Routledge, pp. 295–308.

David, C.–P. and Grondin, D. (eds) (2006) *Hegemony or Empire? The Redefinition of American Power under George W. Bush.* Aldershot: Ashgate.

Dodds, K. (2005) *Global Geopolitics: A Critical Introduction.* Harlow: Pearson.

Dodge, T. (2003) *Inventing Iraq: The Failure of Nation Building and a History Denied.* New York: Columbia University Press.

Farish, M. (2003) 'Disaster and decentralization: American cities and the Cold War', *Cultural Geographies,* 10(2): 125–148.

Farish, M. (2007) 'Targeting the inner landscape', in D. Gregory and A. Pred (eds) *Violent Geographies: Fear, Terror, and Political Violence.* New York: Routledge, pp. 255–271.

Flint, C. (ed.) (2005) *The Geography of War and Peace: From Deathcamps to Diplomats.* Oxford: Oxford University Press.

Fluri, J. (2009) '"Foreign passports only": geographies of (post)conflict work in Kabul, Afghanistan', *Annals of the Association of American Geographers,* 99(5): 986–994.

Foucault, M. (2007) *Security, Territory, Population: Lectures at the Collège de France, 1977–1978* (trans. G. Burchell). Basingstoke: Palgrave Macmillan.

Foucault, M. (2008) *The Birth of Biopolitics: Lectures at the Collège de France, 1978–1979* (trans. G. Burchell). Basingstoke: Palgrave Macmillan.

Gold, D. (1988) *America, the Gulf and Israel: CENTCOM (Central Command) and Emerging US Regional Security Policies in the Middle East,* Jaffee Center for Strategic Studies. Jerusalem: The Jerusalem Post.

Graham, S. (2005) 'Remember Fallujah: demonising place, constructing atrocity', *Environment and Planning D: Society and Space,* 23(1): 1–10.

Gregory, D. (2004) *The Colonial Present: Afghanistan, Palestine, Iraq.* Oxford: Blackwell.

Gregory, D. (2005) 'Geographies, publics and politics', *Progress in Human Geography,* 29(2): 182–193.

Gregory, D. (2007) 'Vanishing points: law, violence and exception in the global war prison', in D. Gregory and A. Pred (eds) *Violent Geographies: Fear, Terror and Political Violence.* New York: Routledge, pp. 205–236.

Gregory, D. and Pred, A. (eds) (2007) *Violent Geographies: Fear, Terror and Political Violence.* New York: Routledge.

Harvey, D. (2003) *The New Imperialism.* Oxford: Oxford University Press.

Harvey, D. (2004) 'Geographical knowledges/political powers', *Proceedings of the British Academy,* 122: 87–115.

Hedges, C. (2003) *War is a Force that Gives us Meaning.* New York: Anchor Books.

Huntington, S.P. (1993) 'The clash of civilizations', *Foreign Affairs*, 72(3): 22–49.
Hyndman, J. (2007) 'Feminist geopolitics revisited: body counts in Iraq', *The Professional Geographer*, 59(1): 35–46.
Jhaveri, N. (2004) 'Petroimperialism: US oil interests and the Iraq War', *Antipode*, 36(1): 2–11.
Johnson, C. (2004) *The Sorrows of Empire: Militarism, Secrecy, and the End of the Republic*. New York: Metropolitan.
Kearns, G. (2006) 'Naturalising empire: echoes of Mackinder for the next American century?', *Geopolitics*, 11(1): 74–98.
Klein, N. (2004) 'Kerry and the gift of impunity', *The Nation*, 13 December.
Klein, N. (2007) *The Shock Doctrine: The Rise of Disaster Capitalism*. New York: Metropolitan.
Morrissey, J. (2009) 'The geoeconomic pivot of the global war on terror: US Central Command and the war in Iraq', in D. Ryan and P. Kiely (eds) *America and Iraq: Policy-Making, Intervention and Regional Politics*. New York: Routledge, pp. 103–122.
Morrissey, J. (2011a) 'Closing the neoliberal gap: risk and regulation in the long war of securitization', *Antipode*, 43(3): 874–900.
Morrissey, J. (2011b) 'Liberal lawfare and biopolitics: US juridical warfare in the war on terror', *Geopolitics*, 16(2): 280–305.
Ó Tuathail, G. (1993) 'The effacement of place: US foreign policy and the spatiality of the Gulf Crisis', *Antipode*, 25(1): 4–31.
Ó Tuathail, G. (1996) *Critical Geopolitics: The Politics of Writing Global Space*. Minneapolis: University of Minnesota Press.

Ó Tuathail, G. (2003) '"Just out looking for a fight": American affect and the invasion of Iraq', *Antipode*, 35(5): 856–870.
Said, E. (1993) *Culture and Imperialism*. New York: Alfred A. Knopf.
Said, E. (1997 [1981]) *Covering Islam: How the Media and the Experts Determine How We See the Rest of the World*. London: Vintage.
Said, E. (2003 [1978]) *Orientalism*. London: Penguin.
Shapiro, M.J. (1997) *Violent Cartographies: Mapping Cultures of War*. Minneapolis: University of Minnesota Press.
Sidaway, J. (1998) 'What is in a Gulf? From the 'arc of crisis' to the Gulf War', in G. Ó Tuathail and S. Dalby (eds) *Rethinking Geopolitics*. London: Routledge, pp. 224–239.
Silberstein, S. (2002) *War of Words: Language, Politics and 9/11*. London: Routledge.
Smith, N. (2003) *American Empire: Roosevelt's Geographer and the Prelude to Globalization*. Berkeley: University of California Press.
Smith, N. (2005) *The Endgame of Globalization*. New York: Routledge.
US CENTCOM, *Shaping the Central Region for the 21st Century*, www.centcom.mil (accessed 20 April 2005).
US Department of Defense (2007) 'U.S. Africa Command. Briefing slide', 7 February, www.defenselink.mil/dodcmsshare/briefingslide/295/070207-D-6570C-001.pdf (accessed 26 June 2013).
US Senate Arms Services Committee (2006) *Statement of General John P. Abizaid, United States Army Commander, United States Central Command, before the Senate Arms Services Committee on the 2006 Posture of the United States Central Command*. 16 March.

Weber, C. (2003) 'The media, the "war on terrorism" and the circulation of non-knowledge', in D.K. Thussu and D. Freedman (eds) *War and the Media: Reporting Conflict 24/7*. London: Sage, pp. 190–199.
Woodward, R. (2004) *Military Geographies*. Oxford: Blackwell.

FURTHER READING

Dalby, S. (2007) 'The Pentagon's new imperial cartography: tabloid realism and the war on terror', in D. Gregory and A. Pred (eds) *Violent Geographies: Fear, Terror and Political Violence*. New York: Routledge, pp. 295–308.
Morrissey, J. (2009) 'The geoeconomic pivot of the global war on terror: US Central Command and the war in Iraq', in D. Ryan and P. Kiely (eds) *America and Iraq: Policy-Making, Intervention and Regional Politics*. New York: Routledge, pp. 103–122.
Ó Tuathail, G. (2003) '"Just out looking for a fight": American affect and the invasion of Iraq', *Antipode*, 35(5): 856–870.
Smith, N. (2003) *American Empire: Roosevelt's Geographer and the Prelude to Globalization*. Berkeley: University of California Press.

Section 3
Historical Hierarchies

7 CLASS, HEGEMONY AND RESISTANCE

Ulf Strohmayer

Introduction

Historically motivated scholarship is not primarily renowned for the conceptual work it bequeaths to future generations; prizing 'accuracy', 'plausibility' and, perhaps, the ability to order developments that take place in some 'longue durée' or another, historical writing in general and historical geography in particular have only relatively recently acquired a more pronounced taste for conceptually centred labour. It is hence no surprise to see concepts that had achieved wide circulation in human geography such as 'class', 'race' or 'gender' – to name but the principal foci of this section of the book – become explicit and legitimized tools of scholarship only towards the latter part of the twentieth century. Of course, the crucial word being 'explicit' because the processes analysed by using such conceptually aware language were always part and parcel of historical writings, if mostly in an implied and somewhat more oblique and adaptable form. The concept of 'class', for instance, might hide in descriptions of (dominant) social groups, while practices associated with 'race' may be subsumed underneath the mantle of less specifically defined historical actors; however, it is precisely in this 'hidden' form that any relevant particular context has been objectified and 'naturalized' in the ensuing historical narratives, whether they were about capitalism, globalism or colonialism (Smith, 2008). The invocation of historically active subjects underneath universally applicable terms like 'mankind', 'nation' or 'culture', for instance, all too often hides or masks more specific motivations, interests or constraints. If nothing else, neglecting to engage with historical actors (or groups thereof) in a more differentiated and nuanced manner would seem to be a wasted opportunity for further and arguably more interesting historical scholarship.

Conversely, the relationship between differentiating categories of the kind articulated in this section and broader historical developments is one of the more interesting areas to study. Arguing from a historically 'deep' perspective, Neil Smith has placed the emergence of differentiated societies within the context of a – historically quite distant – materialization of commodity exchange, and thus of divisions of labour. Key to this development was arguably the emergence of a distinctly 'spatial' element within the organization of historical societies, which increasingly devalued 'place' as the primary organizational principle of pre-modern societies (2008: 107–111). In this reading of historical geography, the rise of a more abstract notion of 'space' facilitates developments that transcend a more narrowly construed set of social and economic relations rooted in 'place'; examples of this would be the rise in trans–local trade relationships, the formation of national territories or, at a smaller scale, the separation of work from residential spaces. Part and parcel of this invalidization of space were increasingly more pronounced social differentiations; 'class' arguably holds pride of place amongst these given its centrality within, as well as for, the different historically dominant production processes.

Geographies With and Without 'Class'

Partly to counteract the tendency to avoid sharper categorical edges, partly in response to more narrowly construed scholarly objectives, some historical geographers began to employ a more conceptually aware and nuanced language from (roughly) the late 1960s onwards. As concerns 'class', its all-too-often pronounced absence from discourses that have shaped the sub-discipline was answered especially by the historically inspired works written by David Harvey (see especially 1985), Derek Gregory's probing paper on the usefulness of class-based historical analyses (1984), and by two edited volumes focusing on the different urban spatialities attaching to class (Thrift and Williams, 1987; Kearns and Withers, 1991). The focus on cities was no accident given the legacy of locating evidence in the built environment that had dominated historical geography for the longest time – a focus that continues in a diminished fashion to the present day (Busteed and Hodgson, 1994; Harris and Sendbuehler, 1994; Hiebert, 1995). Here, differentiations within the built environment, its location, style and social profile were read as evidence of social differentiations or 'class'.

That said, the use of 'class' as an explanatory category in historical geography is still surprisingly underdeveloped given its eminent usefulness and universal applicability. In this respect at least, not much has changed since Mark Billinge's assessment of the situation a generation ago (1982: 23–34). Arguably, the old adage that history is written by the victors not only attaches to national fights for supremacy in wars, but has also been allowed to shape the perception of historical geographies as being classless when more often than not we witness an unfolding of middle class or bourgeois values in the material landscapes we study instead. One reason for the relative neglect especially of 'class' in historical geography stems from a perceived inflexibility attaching to categorical language itself – and to the structure such language provides. Developed largely within sociology and political science, it arguably took the publication of E.P. Thompson's *The Making of the English Working Class* in 1963 to entice a more flexible deployment of 'class' across human geography more generally. Rather than impacting directly on scholarship in historical geography – other than in the context of works mentioned further up – the gain in flexibility has recently led to an incorporation of 'class' into analyses of theoretical narratives relevant to historical geography (Gregory, 1991), heritage (Hardy, 1988) and 'identity' (Proudfoot, 2000). In this respect, most recent work that uses 'class' as a lens for historically motivated inquiries does so in conjuncture or outright amalgamation with other socially differentiating categories, not infrequently resulting in more overtly 'political' readings of both geography and history (Gorlizki, 2000).

From 'Class' to 'Domination' and 'Resistance'

The flexibility imparted to 'class' by the work of Thompson and others is thus very much in evidence in contemporary historical geography. To appreciate the richness of scholarship that has and continues to emerge in the field, we need to remember that in addition to designating spatially anchored social differentiations, the existence of 'classes' also leads to social hierarchies in space. A direct result of the uneven distribution of power, rights and access to resources inherent to, and indeed expressed by, 'class', its effects far exceed the narrower confines of the concept and its linguistic legacy in economically determined discourses. Crucially, such effects also exceed the context created by social relations of production, which have long dominated Marxist analyses of 'class' and 'class relations' (Wright, 1985). Much in the same way that other categories drawn from the Marxist canon have recently been more flexibly deployed (see Brenner, 1997 for a new historical interpretation of 'capital'), 'class' is increasingly analysed as the expression of historical processes that are at once flexible and structured, without leading to or implying uniform interests, forms of consciousness or locations.

It is in this form that 'class' enters into a dialogue with related concepts and processes. Key amongst such aspects associated with both 'class' as a spatially structuring hierarchy and the associated geographies emerging from historically specific divisions of labour are the twin notions of contestation and resistance, building on a recognition of historically relative distributions of power (Revill, 2005). Only partly implied in earlier notions of class struggle (Harvey, 1996: 225), both 'contestation' and 'resistance' have to some extent taken over the mantle never quite appropriated by 'class' in recent human geographic discourses. Historical geography is not an exception; in fact, we can witness a genuine explosion of publications concerned with various historically resonant aspects of resistance, effectively shifting the focus of much scholarship towards a more overt and conscious engagement with questions of power (Harris, 1991).

Centrally involved also in the geographical structurings of both race and gender (Morin and Berg, 2001; Garcia-Ramon, 2003) and thus discussed in more detail in the following two chapters, 'resistance' and 'contestation' as themes in historical geography have been discussed *inter alia* in the context of adaptive cultural practices (Butzer, 2002; Busteed, 2005; Whelan, 2005; Carswell, 2006), national identity (Graham, 1997; Osborne, 1998; Morrissey, 2004), natural resources (Evenden, 2004; Wittayapak, 2008), pictorial forms of representation (Godlewska, 2003), technology as a means of control (Duncan, 2002), occupational practices (Philo, 1998), 'folk geography' (Pred, 1990), 'moral geographies' (Ealham, 2005) and land acquisition (Gruffudd, 1995), as well as the history of geographic thought more broadly construed (McEwan, 1998). The sheer wealth, in terms of both quantity and breadth, of these publications is matched by recent publications in fields of human geography concerned with contemporary issues and is thus not at all exceptional; the ease of its incorporation into the analytical practices that define historical geography are nonetheless somewhat astonishing.

Of particular interest in the present context are recent attempts to analyse historically existent forms of resistance that transgress traditional boundaries and place-bound notions of effectiveness. Featherstone's historical analysis of trans-Atlantic networks of struggles, for instance, aims to overcome a particular problem that has plagued politics based on class consciousness since the days of Marx: the boundedness of experience, information and identity (2005). What furthermore unites much of the historical scholarship currently emerging in the present context is its logical affinity, if not its outright indebtedness, towards the work of Michel Foucault. As will be discussed in more detail in Chapter 22, the pronounced emphasis on issues of discipline and control that permeates Foucault's entire oeuvre has arguably created a space for thinking historical configurations as complex webs of domination and resistance (Mayhew, 2009). Historical geography thus often emerges as *having been* performed (Rose, 2002) and thus perhaps best subjected to genealogical analyses (Matless, 1995). Either way, a majority of the work that has emerged from this vein of analysis is concerned as much with the understanding of historical realities as it is aimed at disrupting the registrars that survive to the present day.

KEY POINTS

- Explicit usage of 'class' in historical geography is restricted to notable exceptions within the literature.
- The absence of 'class' (and of other categorical differentiations derived from the social sciences) has often led to the naturalization of social processes within historical geography.
- However, recent years have seen a surge in publications using 'class' in conjuncture with other socially differentiating categories such as 'gender' and 'race'.
- Class-related and class-produced processes can be deduced from the related practices of 'domination' and 'resistance'.

References

Billinge, M. (1982) 'Reconstructing societies in the past: the collective biography of local communities', in A. Baker and M. Billinge (eds) *Period and Place: Research Methods in Historical Geography*. Cambridge: Cambridge University Press, pp. 19–32.

Brenner, N. (1997) 'Between fixity and motion: accumulation, territorial organization, and the historical geography of spatial scales', *Environment and Planning D: Society and Space*, 16(4): 459–481.

Busteed, M. (2005) 'Parading the green: procession as subaltern resistance in Manchester in 1867', *Political Geography*, 24(8): 903–933.

Busteed, M. and Hodgson, R. (1994) 'Irish migration and settlement in early nineteenth century Manchester, with special reference to the Angel Meadows district in 1851', *Irish Geography*, 27(1): 1–13.

Butzer, K. (2002) 'French wetland agriculture in Atlantic Canada and its European roots: different avenues to historical diffusion', *Annals of the Association of American Geographers*, 82(3): 451–470.

Carswell, G. (2006) 'Multiple historical geographies: responses and resistance to colonial conservation schemes in East Africa', *Journal of Historical Geography*, 32 (2): 398–421.

Duncan, J. (2002) 'Embodying colonialism? Domination and resistance in nineteenth century Ceylonese coffee plantations', *Journal of Historical Geography*, 28(3): 317–338.

Ealham, C. (2005) 'An imagined geography: ideology, urban space, and protest in the creation of Barcelona's "Chinatown", c. 1835–1936', *International Review of Social History*, 50(3): 373–397.

Evenden, M. (2004) 'Social and environmental change at Hell's Gate, British Columbia', *Journal of Historical Geography*, 30(1): 130–153.

Featherstone, D. (2005) 'Atlantic networks, antagonisms and the formation of subaltern political identities', *Social and Cultural Geography*, 6(3): 387–404.

Garcia-Ramon, M.-D. (2003) 'Gender and the colonial encounter in the Arab world: examining women's experiences and narratives', *Environment and Planning D: Society and Space*, 21(6): 653–672.

Godlewska, A. (2003) 'Resisting the cartographic imperative: Giuseppe Bagetti's landscapes of war', *Journal of Historical Geography*, 29(1): 22–50.

Goheen, P. (2003) 'The assertion of middle-class claims to public space in late Victorian Toronto', *Journal of Historical Geography*, 29(1): 73–92.

Gorlizki, Y. (2000) 'Class and nation in the Jewish settlement of Palestine: the case of Merhavia, 1910–30', *Journal of Historical Geography*, 26(4): 572–588.

Graham, B. (ed.) (1997) *In Search of Ireland: a Cultural Geography*. London: Routledge.

Gregory, D. (1984) 'Contours of crisis? Sketches for a geography of class struggle in the Early Industrial Revolution', in D. Gregory and A. Baker (eds) *Explorations in Historical Geography*. Cambridge: Cambridge University Press, pp. 68–117.

Gregory, D. (1991) 'Interventions in the historical geography of modernity: social theory, spatiality and the politics of representation', *Geografiska Annaler*, 73(B)(1): 17–44.

Gruffudd, P. (1995) 'Remaking Wales: nation-building and the geographical imagination, 1925–50', *Political Geography*, 14(3): 219–239.
Hardy, D. (1988) 'Historical Geography and Heritage Studies', *Area* 20(4): 333–338.
Harris, C. (1991) 'Power, modernity and Historical Geography', *Annals of the Association of American Geographers*, 81(4): 671–683.
Harris, R. and Sendbuehler, M. (1994) 'The making of a working-class suburb in Hamilton's East End, 1900–1945', *Journal of Urban History*, 20(4): 486–511.
Harvey, D. (1985) *The Urbanization of Capital: Studies in the History and Theory of Capitalist Urbanization, Volume 2*. Oxford: Blackwell.
Harvey, D. (1996) *Justice, Nature and the Geography of Difference*. Oxford: Blackwell.
Hiebert, D. (1995) 'The social geography of Toronto in 1931: a study of residential differentiation and social structure', *Journal of Historical Geography*, 21(1): 55–74.
Kearns, G. and Withers, C. (eds) (1991) *Urbanising Britain: Essays on Class and Community in the Nineteenth Century*. Cambridge: Cambridge University Press.
Matless, D. (1995) 'Effects of history', *Transactions of the Institute of British Geographers*, 20(4): 405–409.
Mayhew, R. (2009) 'Historical Geography 2007–08: Foucault's avatars – still in (the) Driver's seat', *Progress in Human Geography*, 33(3): 387–397.
McEwan, C. (1998) 'Cutting power lines within the palace? Countering paternity and Eurocentrism in the "geographical tradition"', *Transactions of the Institute of British Geographers*, 23(3): 371–384.
Morin, K. and Berg, L. (2001) 'Gendering resistance: British colonial narratives of wartime New Zealand', *Journal of Historical Geography*, 27(2): 196–222.
Morrissey, J. (2004) 'Geography militant: resistance and the essentialisation of identity in Colonial Ireland', *Irish Geography*, 37(2): 166–176.
Osborne, B. (1998) 'Constructing landscapes of power: the George Etienne Cartier monument in Montreal', *Journal of Historical Geography*, 24(4): 431–458.
Philo, C. (1998) 'A "lyffe in pyttes and caves": exclusionary geographies of the West Country tinners', *Geoforum*, 29(2): 159–172.
Pred, A. (1990) *Lost Words and Lost Worlds: Modernity and the Language of Everyday Life in Later Nineteenth Century Stockholm*. Cambridge: Cambridge University Press.
Proudfoot, L. (2000) 'Hybrid space? Self and Other in narratives of landownership in nineteenth century Ireland', *Journal of Historical Geography*, 26(2): 203–221.
Revill, G. (2005) 'Railway labour and the geography of collective bargaining: the Midland Railway strikes of 1879 and 1887', *Journal of Historical Geography*, 31(1): 17–40.
Rose, M. (2002) 'The seductions of resistance: power, politics, and a performative style of systems', *Environment and Planning D: Society and Space*, 20: 383–400.
Smith, N. (2008) 'What happened to class?', *Environment and Planning A*, 32: 1011–1031.
Thompson, E.P. (1963) *The Making of the English Working Class*. London, V. Gollancz.
Thrift, N. and Williams, P. (eds) (1987) *Class and Space: the Making of Urban Society*. London: Routledge.
Whelan, Y. (2005) 'Performing power, demonstrating resistance: interpreting Queen Victoria's visit to Dublin in 1900', in L. Proudfoot. and M. Roche (eds) *(Dis-)placing Empire: Renegotiating British Colonial Geographies*. Farnham: Ashgate, pp. 99–116.

Wittayapak, C. (2008) 'History and geography of identifications related to resource conflicts and ethnic violence in Northern Thailand', *Asia Pacific Viewpoint*, 49(1): 111–127.
Wright, E.O. (1985) *Classes*. London: Verso.

FURTHER READING

Billinge, M. (1982) 'Reconstructing societies in the past: the collective biography of local communities', in A. Baker and M. Billinge (eds) *Period and Place: Research Methods in Historical Geography*. Cambridge: Cambridge University Press, pp. 19–32.
Duncan, J. (2002) 'Embodying colonialism? Domination and resistance in nineteenth century Ceylonese coffee plantations', *Journal of Historical Geography*, 28(3): 317–338.
Gregory, D. (1984) 'Contours of crisis? Sketches for a geography of class struggle in the Early Industrial Revolution', in D. Gregory and A. Baker (eds) *Explorations in Historical Geography*. Cambridge: Cambridge University Press, pp. 68–117.

8 RACE

David Nally

Introduction

In 1800 European powers claimed ownership over approximately 35 per cent of the planet's surface. Just prior to the great 'scramble for Africa' that proportion had increased to 67 per cent, a rate of increase of 83,000 square miles per annum. By the eve of World War I, the annual rate had risen to 240,000 square miles with Europe then owning 'a grand total of roughly 85 per cent of the earth as colonies, protectorates, dependencies, dominions, and commonwealths' (Said, 1993: 8). 'Globalization by force', as historian Niall Ferguson (cited in Bowles, 2007: 94) describes this period, turned Europeans into self-described masters of the universe (see also Chapter 19).

In his essay 'Geography and Some Explorers,' published in 1924, novelist and critic Joseph Conrad described the consolidation of European empire as 'the vilest scramble for loot that ever disfigured the history of human conscience and geographical exploration' (1999: 27). What makes imperial violence and suffering seem acceptable, and even necessary, Conrad believed, was the perfection of a legitimating apparatus, a set of ideas and theories that invalidate critique, making protest and dissent seem unnecessary or futile. This point about the ideology of imperialism is made with great effect in Conrad's powerful novella *Heart of Darkness* (first serialized in 1899) in which the narrator Marlow says:

> The conquest of the earth, which mostly means the taking it away from those who have a different complexion or slightly flatter noses than ourselves, it not a pretty thing when you look into it too much. What redeems it is the idea only. An idea at the back of it; not a sentimental pretence but an idea and an unselfish belief in the idea – something you can set up, and bow down before, and offer a sacrifice to ... (1999: 65–66)

It is indeed significant that at the same time that Europeans were dividing the world amongst themselves, geography was emerging as a fully-fledged academic discipline in schools and universities, and, as Nayak and Jeffrey (2011: 5) observe, much evidence suggests that 'the ideas and theories that geographers were developing help justify the process of imperial expansion'. Indeed from its very inception, as Kobayashi and Peake (2000: 399) claim, geography was 'a discipline founded … on difference and hierarchy' and racial theories, in particular, were accorded scientific status by very many geographers. As Marlow might have said, the idea of race helped 'redeem' imperialism; it formed a part of a validating discourse that imperialists could 'bow down before, and offer a sacrifice to'.

The Idea of Race

Anthropologist Nicholas Thomas (1994: 79) suggests that race

> ought to be seen as discourse that engages in conceptual and perceptual government, in its apprehension and legislation of types, distinctions, criteria for assessing proximity and distance, and in its more technical applications – in, for instance, notions stipulating that certain forms of labour are appropriate to one race but not another.

Thomas is emphasizing the tremendous 'cultural work' involved in making racial distinctions – 'difference,' he says, 'is produced, not simply distorted' (1994: 205). Before the advent of 'scientific racism' explorers, travel-writers, artists and scholars produced accounts that accentuated the moral, cultural and physiological differences between peoples and regions. Geographer David Livingstone has drawn attention to the way that climatic conditions were used to partition global space 'into a series of bi-polar antipathies: the salubrious and insalubrious; the superior and the subordinate; the moderate and the intemperate' (2002: 173). Inhabitants of the so-called 'Tropics', for example, were judged to be slothful, feeble and licentious, in comparison to their sturdy, industrious and morally chaste cousins living in the more temperate regions of the planet. The renowned philosopher and essayist David Hume (1711–1776) thought that 'the genial heat of the sun, in the countries exposed to his beams, inflames the blood, and exalts the passion between the sexes' (cited in Livingstone 2002: 163). The demographer and economist Thomas Robert Malthus (1776–1834) took up this point in his celebrated *Essay on the Population Principle* (first published in 1798). Malthus believed that the poor, especially the non-European poor, were creatures of nature who bred with little consideration of the consequences. Malthus believed that in the 'southern countries', where the 'passion between the sexes' operated unchecked and the inhabitants lived 'in so degraded state', only the threats of war, pestilence and famine could suppress additional population surges (1989: 147). This racial interpretation of climatic conditions had an enormous bearing on imperial policies in India, for example, where some officials viewed subsistence crises and other 'natural' calamities as means of inculcating moral habits amongst a reluctant and improvident people (Davis, 2001).

The statements by Hume and Malthus imply that a person's moral character 'derived from climatic circumstances', and 'that nature governed, to some degree at least, the global pattern of civilization' (Livingstone, 2002: 168). But as Livingstone (2002: 168) stresses, this brand of 'moral climatology' existed alongside other similar arguments claiming that moral qualities were not derived but '*distributed* according to climatic zone'. This distinction was important, for the latter set of theories and ideas implied that white colonial masters who had migrated to the 'Tropics' might in time succumb to local climate conditions and degenerate into a state of moral and physical torpor – a condition derisively referred to by Europeans as 'going native'. To avoid such perils Europeans added various kinds of filters, both physical and cultural, to limit their 'exposure' to uncontrollable elements. Such 'filters' created a sense of imperial fraternity amongst the colonists and allowed them to maintain the fiction of sovereignty over local environments as much as over local peoples. This is a key argument in James Duncan's (2007) study of the plantation economy in Ceylon, *In the Shadows of the Tropics*. The 'ideal of the powerful European male', explains Duncan (2007: 49), was maintained by forcing colonial administrators to retire at the age of 55 so that 'no Oriental was ever allowed to see a Westerner as he aged and degenerated, just as no Westerner needed to ever see himself, mirrored in the eyes of the subject race, as anything but a vigorous, rational, ever-Alert young, Raj' (Said, cited in Duncan, 2007: 49). 'It was considered absolutely necessary', concludes Duncan (2007: 49), 'for the British to remain in control of themselves to serve as models for the Ceylonese.' (see also Chapters 1 and 2)

David Lambert (2005) takes up this theme in his important study of racial politics in colonial Barbados. Set against the backdrop of the abolition movement, Lambert explores what it meant to be a white colonial subject at a time when the slave trade and the very idea of chattel slavery was under relentless metropolitan attack. In Britain reformers worried that mass enslavement was incompatible with metropolitan conceptions of liberty and justice. In contrast slave owners in the West Indies saw this as an assault not only on their livelihoods, but also on their understanding of themselves as loyal English subjects. The spatial imaginary that distinguished the slave-owning colonies from the 'free' world also served to denigrate the West Indian colonists, whose desire to retain the institution of slavery was seen as proof of their cruel and amoral natures. This mapping of difference did not go uncontested, however, as planters sought to assert the economic value and strategic importance of the colonies as well as their own cultural affinity with Britain. Far from being 'un-English' Barbados was represented as the epitome of Anglo-Saxon values, a 'little England' in the Caribbean and a bulwark against Revolutionary America (Lambert, 2005: 13–15). Lambert successfully shows that at stake in this discursive confrontation was the very concept of a shared 'whiteness'. '[T]he age of abolition', he (2005: 16) explains, 'was, in part, a "war of representation" over "good" and "bad" forms of white identity and the imaginative location of these in the British Atlantic world.' The point that certain groups can be depicted as 'white but not quite' – to paraphrase Homi Bhabha (1994: 89) – will be taken up again later in the chapter.

The idea that moral qualities can be assigned a geographical address – that 'we' over 'here' are fundamentally different from 'them' over 'there' – is sometimes described as *environmental determinism*. Environmental determinism rose to prominence in the late nineteenth and early twentieth century before it was roundly discredited by the majority of social scientists (but see Robert Kaplan's *The Revenge of Geography* (2012) for a recent revival of this kind of thinking). In the United States the geographer Ellen Churchill Semple (1863–1932), helped popularize the belief that character and culture could be explained by differences in the physical environment. Semple had been greatly influenced by the German geographer Friedrich Ratzel (1844–1904) who fused environmental determinism with theories of heritability – derived from Jean-Baptiste Lamarck (1744–1829) – to develop a biological conception of geography and an organic conception of the state. According to Ratzel, states are akin to natural organisms that grow and expand with time; stronger states will naturally usurp space from weaker ones and *ipso facto* the demise of one nation or state is the necessary condition for the healthy expression of another (Agnew and Corbridge 1995). Ratzel's notion of *lebensraum* (meaning 'living space') was adopted by the Nazi regime and used to justify expansionary wars and genocide as a first step to ensuring the healthy future of the German state and people.

Deconstructing Race

Three points are worth emphasizing at this juncture. First, although in modern times it is common to see the extermination of European Jews as an 'anti-western aberration' (Traverso, 2003) – and an episode of absolute evil (Eagleton, 2010) – in fact it can be been shown that the extreme chauvinism and biological racism of the Nazis was profoundly connected to the culture and practice of imperialism that had enveloped Europe since the beginning of the nineteenth century. The connection between fascism and imperialism was first proposed by Hannah Arendt (1976) – when she referred to the 'boomerang effect' of imperial power – and has since been developed and elaborated by Michel Foucault (1980, 2003), Giorgio Agamben (1995, 2002), Jürgen Zimmerer (2005) and Enzo Traverso (2003). For these different writers the attempt to exterminate the Jews (and indeed other 'social degenerates' including homosexuals, Roma people, and the disabled) owed a great deal to the forms of racial persecution practised by Europeans in the process of acquiring and maintaining imperial possessions. Benjamin Madley (2005), for example, has argued that many facets of German colonial policy in South-West Africa – including the employment of slave labour, the building and use of concentration camps (the first German concentration camps were built in colonial Namibia in 1904), the acceptance of the view that the colonized were sub-human, the implementation of *lebensraum* theory and the adoption of *Vernichtung* (or annihilation) during military confrontations – foreshadowed in ominous ways the programme of race murder unleashed by Nazis during World War II.

Second, it should be stressed that the belief in racial purity and a hierarchy of human groups was less controversial and more roundly endorsed in European and American culture than is commonly acknowledged. The rise of Social Darwinism, the study of heredity and the science of eugenics, the statistical measurement of human variance (psychometrics, craniology, etc.) and the birth of racial anthropology show that the idea of race was internalized within modern culture in a wide variety of ways. Furthermore in the United States of America – a country which had of course helped to liberate Europe from Nazi rule – forced sterilization was practised on sections of its prison population and a system of racial segregation was enforced through legislation and extra-legal violence, until the civil rights movement in the 1960s forced the equal recognition of African-Americans. In South Africa racial segregation begun in colonial times under Dutch rule was preserved through Apartheid legislation until the 1990s. These overt examples of racialized discrimination exist alongside less extreme or more everyday patterns of prejudice that often escape critical scrutiny and censure. Indeed, for geographer Ash Amin (2012) racism takes both explicit and banal forms. Accumulations of race are often 'latent', he warns, ready to erupt in the present in forms that may or may not draw on the anti-racist practice of the past.

Finally, and related to Amin's point about banal and latent racisms, it is important to stress that race – like gender, sexuality and class (see also Chapters 7 and 9) – is neither an autonomous discourse nor a free-floating classificatory apparatus, but something that takes its meaning from its conjunction with other tropes and motifs (Greenblatt, 1991). Discussing the representation of the Irish in the nineteenth century, for example, cultural scholar Anne McClintock (1995: 53) argues that English racism drew upon a *gendered* 'notion of the domestic barbarism ... as a marker of racial difference'. Where skin colour proved 'imprecise and inadequate', domestic disorder and personal hygiene could be seized upon as a positive sign of Irish barbarism. The Irish might be white, but they were still fundamentally 'un-English' – and this could be *shown* by their habits of dress, the disarray of their cabins, the accumulations of filth and by the presence of animals living cheek-by-jowl with their owners (Lebow, 1976; Nally, 2011). Before racial science posited a 'Celtic race' and distinguished it from its 'Anglo-Saxon' cousin (Curtis, 1997), the gendered image of domestic vice marked the Irish as irredeemably 'Other' and paved the way for external intervention. British rule was thus represented as a benevolent act, a moment of rescue and redemption (see also Chapter 2).

McClintock's historical narrative reminds us that race is much more than skin colour and that whiteness, despite its privileged status, is also subject to racial profiling. Indeed, 'whiteness studies' is now a burgeoning field of academic enquiry. Geographers such as Alastair Bonnett (2000) and Daniel Swanton (2010) have shown that 'metaphysical and moral hierarchies' (Alcoff, cited in Swanton 2010: 460) are frequently used to differentiate between different sub-categories of 'white'. The common use of the pejorative moniker 'chav' to describe working-class youth culture in the United Kingdom – and other slurs such as 'trailer-trash', 'pikey' and 'redneck' directed at white communities deemed to be of low social status – illustrates how radical difference is insinuated in everyday speech acts. Swanton's work goes further in recording how race is a complex *mediated practice*. Seen this way, traditions, customs, food habits, communication practices (ranging from interpersonal relations to virtual and web-based forms of interaction), clothing and religious apparel, non-human objects and attachments (the car you drive, the kind of CDs you own, etc.) – all mesh to form an approximation of difference. 'Differentiation', writes Swanton (2010: 460), 'is at least as much about relations between bodies, things, and spaces as it is about discourse.' For Amin (2012: 26) mediated practices are ways to embed oneself in relations with others; they are 'part of the sorting machine of good and bad citizens'.

Epistemologies of Difference

The goal of much of the academic work just discussed is to de-naturalize and de-ontologize race by illustrating its inherent constructedness (Graham, 2000; Nally, 2009). Notwithstanding this laudable goal race continues to endure. In a recent trip to Israel, for example, the Republican candidate for the Presidency of the United States, Mitt Romney, attracted negative attention for his suggestion that the differences in wealth between Israel and Palestine could be explained by a combination of 'geographic elements' (Cassidy, 2012) and cultural values (interestingly Romney's speech drew on Jared Diamond's (1997) book *Gun, Germs and Steel*, which has been roundly criticized for revivifying the language of environmental determinism to explain the successes and failures of 'civilizations' (cf. Diamond, 2012)). Note that Romney did *not* have to use the word 'race' in his speech for his (largely Israeli) audience to pick up on his suggestion that Palestinian poverty was caused not by the military garrisoning carried out by an occupying

Israeli army, nor by the trade restrictions enforced by the Israeli state, but by the failings of Palestinian culture itself. According to this logic, poverty is causally linked to a *pathological population* (Foucault 2004). Here race hovers as a 'sorting machine', even though the word is never mentioned in Romney's address.

Race endures in other ways too. During the BBC's coverage of the 2012 London Olympics the television pundit John Inverdale asked his studio analysts, Denise Lewis, Michael Johnson and Colin Jackson – all black, former athletes – why it is that 81 of the 82 men who have run the 100 metres in under ten seconds are black. Following Inverdale's question a short documentary clip was shown. The documentary asked, among other things, whether the historical institution of chattel slavery heightened 'natural selection' pressures giving descendants of slaves an 'advantage' in physical events like athletics. In the subsequent studio discussion, Michael Johnson – who had previously participated in the documentary film *Survival of the Fastest* broadcast on Channel 4 in July 2012 – challenged the implied connection between the genes of surviving slaves and the prowess of black athletes in modern sprint events. Johnson pointed out that funding, training facilities, personal determination and grit, as well as social forces that promote one sport over another, are far more convincing explanations than race for the success of black athletes in track and field events.

The BBC's clip and Channel 4's documentary film are in one sense an expression of renewed public interest in race as articulated through the lens of genomic science (witness the popularity of television programmes like *Crime Scene Investigation* and the 'Geneographic Project', sponsored by the National Geographic Society and IBM). Ironically genomic science has 'confirmed that some 98 per cent of the human genetic pool is shared with chimpanzees, and that variations in DNA sequence are greater within than between human groups (Amin, 2012: 85–86; cf. Rose, 2007: 168). 'Yet', as Amin (2012: 86) continues, 'the very science that questions race as a marker of human difference is being mobilized to validate physical or behavioural differences': obesity, heart disease, educational attainment and criminality (e.g. the Bush administration's so-called *Violent Initiative* designed to identify people early in life who are 'prone' to anti-social behaviour) are all problems that have been causally linked to genetic predispositions. The classificatory logic that supports an insurance company's decision to deny medical cover to population groups deemed to be 'at risk' to certain diseases is the self-same logic that attributes a sprinter's performance to their racial background. It is critically important to recognize that genetic determinism is no less corrosive for implying that 'black people are faster' than for insisting that 'black people are inferior'.[1]

At its most basic level race is an epistemology of difference, a means calibrating the strangeness of others, whilst reinforcing a more positive sense of oneself. We have seen that the history of 'race' as an organizing concept is not the same as the history of 'racism' as an overt political programme. The theory of 'moral climatology' and the notion of 'environmental determinism' transformed what David Theo Goldberg (2000: 155) terms 'the spaces of the Other' (colonies, plantations, modern ganglands and urban ghettoes) into social laboratories in which the salience of race as an 'epistemological construct' was tested and refined (cf. Santiago-Valles 2006). A geographical imaginary played a primary role in establishing sharp caesuras between human populations. Over and over again distance is folded into distance creating what Derek Gregory (2004) refers to as an 'architecture of enmity'. The architectural metaphor is appropriate for at heart *these categorisations are constructs* – they are 'mental landscapes' that tell us much more about the fears and anxieties of those who find strangeness intolerable than they do about the implied difference of one group from another. These 'essentialized' constructions of human difference were the building blocks for the biological racism employed by the Nazis, and they are resonant in the forms of social differentiation practised in the present: the casting of Muslims as 'terrorists', the xenophobia surrounding emigration, popular aversion toward asylum seekers, hysterical warnings about cultural dilution (present in calls to implement 'nationality tests', for example), the branding of welfare claimants as good-for-nothings, and the routine discrimination levelled at travellers and Roma peoples (cf. Amin, 2012).

Conclusion

Philosopher Tzvetan Todorov (2000: 64) has claimed the 'the ordinary racist is not a theoretician'. While this may be the case, Amartya Sen's (2006: xvi) contention that 'the reductionism of high theory can make a contribution, often inadvertently, to the violence of low politics' is surely borne out in the sense of threat that routinely attends the apprehension of difference. The recent violence unleashed by Anders Behring Breivik in Norway and Wade Michael Page in Wisconsin is illustrative in this regard. Breivik and Page are not usually described as 'theoreticians', although significantly both men took up violence following a long immersion in far-right thought encountered in online 'virtual communities', in Breivik's case, and in the 'white power rock' music scene of which Page was an active member (Goodwin, 2012). Racism is indeed a kind of 'theoretical disposition', a form of 'travelling theory', to borrow Edward Said's (1983) term, that passes from person to person and from situation to situation, taking on different allusions and meanings depending on where, when and how it is deployed. And it is for this very reason that the discipline of Geography, tarnished by its association with imperial wars of the past, can play a vital role in transcending racism in the present.

KEY POINTS

- Historical geographers have show how racial thinking emerged from prior epistemologies of difference, including moral climatology and environmental determinism.
- Racism was central to colonial and later imperial expansion. The discourse of race sanctioned deeper and more intrusive modes of metropolitan supervision. It made the unequal treatment of indigenous populations seem necessary.
- Historical geographers have shown how the idea of race takes on different meanings and allusions depending on where, when and how it is deployed. Racism can be said to develop spatially.
- Racism can take both explicit and banal forms. Everyday speech acts often give expression to racist practices of the past.

Note

1 An astute point made by a blogger, David King, in response to the documentary film *Survival of the Fastest*. See www.channel4.com/programmes/michael-johnson-survival-of-the-fastest/episode-guide/series-1/episode-1#comments-top (accessed August 2012).

References

Agamben, G. (1995) *Homo Sacer: Sovereign Power and Bare Life* (trans. D. Heller-Roazen). Stanford, CA: Stanford University Press.

Agamben, G. (2002) *Remnants of Auschwitz: The Witness and the Archive* (trans. D. Heller-Roazen). New York: Zone Books.

Agnew, J. and Corbridge, S. (1995) *Mastering Space: Hegemony, Territory and International Political Economy*. London: Routledge.

Amin, A. (2012) *Land of Strangers*. Cambridge: Polity Press.

Arendt, H. (1976) *The Origins of Totalitarianism*. New York: Harcourt.

Bonnett, A. (2000) *White Identities: Historical and International Perspectives*. Harlow: Prentice Hall.

Bhabha, H. (1994) *The Location of Culture*. London: Routledge.

Bowles, P. (2007) *Capitalism*. Harlow: Pearson.

Cassidy, J. (2012) 'Memo to Mitt: The Palestinians' problems aren't all cultural', *New Yorker*, 30 July. www.newyorker.com/online/blogs/johncassidy/2012/07/memo-to-mitt-the-palestinians-problems-arent-all-cultural.html#ixzz2A9iA09Lj (accessed 1 October 2012).

Conrad, J. (1999) *Heart of Darkness* (ed. D.C.R.A Goonetilleke, 2nd edn). Peterborough, ON: Broadview Literary Texts.

Curtis Jr., L.P. (1997) *Apes and Angels: The Irishman in Victorian Caricature*. London: Smithsonian Institution Press.

Davis, M. (2001) *Late Victorian Holocausts: El Niño Famines and the Making of the Third World*. London: Verso.

Diamond, J. (1997) *Guns, Germs, and Steel: The Fates of Human Societies*. New York: Norton.

Diamond, J. (2012) 'Romney hasn't done his homework', *New York Times*, 1 August. www.nytimes.com/2012/08/02/opinion/mitt-romneys-search-for-simple-answers.html (accessed 1 October 2012).

Duncan, J.S. (2007) *In the Shadows of the Tropics: Climate, Race and Biopower in Nineteenth Century Ceylon*. Aldershot: Ashgate.

Eagleton, T. (2010) *On Evil*. London: Yale University Press.

Foucault, M. (1980) *The History of Sexuality. Vol. 1: An Introduction* (trans. R. Hurley). New York: Vintage Books.

Foucault, M. (2003) *Society Must Be Defended: Lectures at the Collège de France, 1975–1976* (ed. Arnold I. Davidson, trans. D. Macey). New York: Picador.

Foucault, M. (2004) *Abnormal: Lectures at the Collège de France 1974–1975* (ed. Arnold I. Davidson, trans. G. Burchell). New York: Picador.

Goldberg, D.T. (2000) 'Racial knowledge', in Les Black and John Solomos (eds) *Theories of Race and Racism: A Reader*. London: Routledge, pp. 154–180.

Goodwin, M. (2012) 'Wade Michael Page and the rise of violent far-right extremism', *Guardian*, 8 August. www.guardian.co.uk/world/2012/aug/08/wade-michael-page-violent-far-right (accessed 1 October 2012).

Graham, B. (2000) 'The past in place: historical geographies of identity', in B. Graham and C. Nash (eds) *Modern Historical Geographies*. London: Pearson, pp. 70–99.

Greenblatt, S. (1991) *Marvelous Possessions: The Wonder of the New World*. Chicago: University of Chicago Press.

Gregory, D. (2004) *The Colonial Present: Afghanistan, Palestine, Iraq*. Oxford: Blackwell.

Kaplan, R. (2012) *The Revenge of Geography: What the Map Tells Us about Coming Conflicts and the Battle Against Fate*. New York: Random House.

Kobayashi, A. and Peake, L. (2000) 'Racism out of place: thoughts on whiteness and an antiracist geography in the new millennium', *Annals of the Association of American Geographers*, 90(2): 392–403.

Lambert, D. (2005) *White Creole Culture, Politics and Identity during the Age of Abolition*. Cambridge: Cambridge University Press.

Lebow, N. (1976) *White Britain and Black Ireland: The Influence of Racial Stereotypes on Colonial Policy*. Philadelphia: ISHI.

Livingstone, D.N. (2002) 'Race, Space and Moral Climatology: Notes Toward a Genealogy', *Journal of Historical Geography*, 28(2): 159–180.

Madley, B. (2005) 'From Africa to Auschwitz: how German South West Africa incubated ideas and methods adopted and developed by the Nazis in Eastern Europe', *European History Quarterly*, 35(3): 429–464.

Malthus, T.R. (1989) *An Essay on the Principle of Population Volume II* (ed. Patricia James). Cambridge: Cambridge University Press.

McClintock, A. (1995) *Imperial Leather: Race, Gender and Sexuality in the Colonial Context*. London: Routledge.

Nally, D. (2009) 'Historical Geographies of Ethnicity and Resistance', in Rob Kitchin and Nigel Thrift (eds) *International Encyclopedia of Human Geography, Volume 3*. Oxford: Elsevier, pp. 620–625.

Nally, D. (2011) *Human Encumbrances: Political Violence and the Great Irish Famine*, Notre Dame, IN: University of Notre Dame Press.

Nayak, A. and Jeffrey, A. (2011) *An Introduction to Geographical Thought*. Harlow: Pearson.

Rose, N. (2007) *The Politics of Life Itself: Biomedicine, Power, and Subjectivity in the Twentieth-First Century*. Princeton, NJ: Princeton University Press.

Said, E. (1983) *The World, the Text and the Critic*. Cambridge, MA: Harvard University Press.

Said, E. (1993) *Culture and Imperialism*. New York: Vintage Books.

Santiago-Valles, K. (2006) '"Bloody legislations", "entombment" and race making in the Spanish Atlantic: differentiated spaces of general(ized) confinement in Spain and Puerto Rico, 1750–1840', *Radical History Review*, 96: 33–57.

Sen, A. (2006) *Identity and Violence: The Illusion of Destiny*. New York: W.W. Norton and Company.

Swanton, D. (2010) 'Flesh, metal, road: tracing the machinic geographies of race', *Environment and Planning D: Society and Space*, 28(3): 447–466.

Thomas, N. (1994) *Colonialism's Culture: Anthropology, Travel and Government*. Princeton, NJ: Princeton University Press.

Todorov, T. (2000) 'Race and racism', in Les Black and John Solomos (eds) *Theories of Race and Racism: A Reader*. London: Routledge, pp. 64–70.

Traverso, E. (2003) *The Origins of Nazi Violence* (trans. J. Lloyd). New York: New Press.

Zimmerer, J. (2005) 'Annihilation in Africa: the "race war" in German Southwest Africa (1904–1908) and its significance for a global history of genocide', *Bulletin of the German Historical Institute*, 37 (Fall): 51–57.

FURTHER READING

Amin, A. (2012) *Land of Strangers*. Cambridge: Polity Press.
Duncan, J.S. (2007) *In the Shadows of the Tropics: Climate, Race and Biopower in Nineteenth Century Ceylon*. Aldershot: Ashgate.
Lambert, D. (2005) *White Creole Culture, Politics and Identity during the Age of Abolition*. Cambridge: Cambridge University Press.
Livingstone, D.N. (2002) 'Race, space and moral climatology: notes toward a genealogy', *Journal of Historical Geography*, 28(2): 159–180.

9 GENDER

David Nally

Introduction

In December 2012, during one of his most important speeches of the year, Pope Benedict XVI, leader of the Roman Catholic Church, the world's largest Christian organization, announced that a person's sexuality and gender identity is God-given and unalterable. 'People dispute the idea that they have a nature, given to them by their bodily identity, that serves as a defining element of the human being', the Pope declared before a rapt Vatican audience. 'They deny their nature and decide that it is not something previously given to them, but that they make it for themselves.' As it turns out the Pope's admonition was directed at gay couples and activists whose efforts to legalize homosexual marriage constituted what he described as a grave threat to human dignity. 'When the freedom to be creative becomes the freedom to create oneself', Pope Benedict concluded, 'then necessarily the Maker himself is denied and ultimately man too is stripped of his dignity as a creature of God, as the image of God at the core of his being' (Pope Benedict XVI, 2012).

Pope Benedict's remarks outline the stakes involved in what is sometimes summarized as the 'nature–nurture' debate (see also Chapter 21). In denying gay couples the right to marry the Catholic Church is attempting to erase from public space what it considers to be deviant forms of bodily identity. The first step in this act of erasure is to insist that homosexuality is a 'sin against nature' – an opinion expressed recently by several Anglican bishops and a timely reminder that such views are not confined to a single church or denomination. In fact the arguments go beyond an attack on gay couples and activists. For at the heart of this debate is whether a person's bodily identity is 'natural' or socially and historically produced (Duncan 1996a). The Catholic Church's belief that a person's gender and sexual identity is decided by God – the ultimate 'author of life' – might be usefully juxtaposed with the views of Florentino Ariza, the fictional character in Gabriel García Márquez's (1988: 165) magnificent novel *Love in the Time of Cholera*, who 'allowed himself to be swayed by his conviction that human beings are not born once and for all on the day their mothers give birth to them, but that life obliges them over and over again to give birth to themselves'.

The Social Construction of Gender

The assertion that human beings can and do 'give birth to themselves' – that biology is not destiny – is now a common refrain in the humanities and social sciences. In her justly renowned book *The Second Sex*, first published in 1949, and incidentally placed on a list of forbidden books by the Vatican (du Plessix Gray, 2010), French intellectual and activist Simone de Beauvoir claimed that 'one is not born, but rather becomes, a woman' (1989: 267). De Beauvoir was insisting that differences between men and women – between 'masculinity' and 'femininity' – are attributable to cultural processes rather than biological facts. Historically in most societies, she pointed out, inheritance laws prioritized the succession rights of men ensuring that women remained propertyless and therefore economically dependent on males. In public institutions women are socialized to accept gender-based distinctions (think of single-gender schools or 'male' and 'female' wards in hospitals) and submit to patriarchal norms and values (consider the social pressures to adopt particular vocational roles for 'women' and 'men'). Later the responsibility to bear and raise children confines women to the domestic sphere, while men are free to pursue public careers and remunerated work. The entire social world bears the deep impress of gender divisions – even the writing of the past is enshrined as 'his-story'. In a multitude of ways, de Beauvoir claimed, women are historically and socially assimilated into a position of subservience vis-à-vis men (cf. Butler, 1986).

The Second Sex is commonly described as a foundational text and harbinger of the 'second wave' of feminist thought. Of course it built on an earlier history of feminist criticism – the so-called 'first wave' that included now classic texts such as Mary Wollstonecraft's *A Vindication of the Rights of Woman* (1972; first published in 1792) and John Stuart Mill's *The Subjection of Women* (first published in 1869 and tellingly

Mill's only book to lose his publisher money) – as well as popular movements, such as the suffragettes, who militantly campaigned for women's voting rights. What distinguishes the second wave from its nineteenth-century progenitor, however, is the shift in focus from an exploration of unequal gender roles to studying the nexus of power relations that undergird the construction of gender itself (Mitchell, 2000). The liberatory project was now, in Gibson-Graham's (1994: 219) words, 'to undermine the hegemony of the binary gender discourse and to promote alternative subject positions for gendered subjects'.

Historicizing the Sex–gender Distinction

Beginning in the 1970s 'third wave' feminism, heavily influenced by European post-structuralist thought, sought to challenge the 'sex–gender distinction' left 'analytically untouched' by the second wave, according to the influential feminist critic Donna Haraway. Although it was more and more acknowledged that gender was 'cultural', sex was still read as the 'essence' of the human subject and thus the 'core' of a person's gender identity. 'The root difficulty was an inability to *historicize* sex itself', wrote Haraway (1991: 131). This difficulty was compounded by the fact that feminists frequently made use of the 'sex differences' argument to highlight cases of social exclusion. Indeed without accepting an essentialised formulation of identity (as 'woman' or 'man') it would be very difficult to make a case for sexual discrimination. For Haraway this form of 'tactical usefulness' (Haraway, 1991: 136) – what Spivak (1987: 205) earlier termed 'strategic essentialism' – carries the immense cost of constructing lesbian, gay, bisexual, transsexual and transgendered individuals as outside and deviant from the normative 'male' and 'female' modes of identification. As Haraway points out, 'sex' does not pre-exist its construction (Kobayashi, 2001) – 'biology is a discourse, not the living world itself' (Haraway, 1992: 298).

For geographers Bondi and Davidson (2003) second wave feminism also left untroubled the conceptual separation of subjecthood and environment, such that everyday spaces often conjure up sharp gender associations. 'Female boxers and brickies are likely to attract comment as unwomanly or "mannish"', they write, 'while male flight attendants and secretaries are often characterised as effeminate or effete. Precisely because these various places bear such powerful gender inscriptions, men and women put their masculine or feminine credentials in question through their mere presence in those environments' (Bondi and Davidson, 2003: 328–329). Here the 'anti-foundationalist' position elaborated by authors such as Donna Haraway and Judith Butler is extended to include the treatment of space. Far from being distinct spheres of experience and social activity, sex, gender and space are reciprocally constituted at different scales and with different intensities of interaction. 'To *be*', as Bondi and Davidson (2003: 338) memorably put it, is necessarily 'to be some*where*.'

Gendering Historical Geography

As the latter remarks attest, feminist criticism is today a vibrant strand of critical human geography. Feminist scholars have shaken up not only the history of the discipline, but also the conduct of historical geography. Gillian Rose (1993), for example, has subjected the history of geographic thought to a feminist critique. According to Rose, Geography, like many other subjects in the social and physical sciences, has traditionally been thought of as a male enterprise. In Mona Domosh's (1991: 102) words, 'gender relations and representations are integral to the social construction of knowledge' – geographical knowledge included. However, rather than question the gendered norms that mottle the subject's early development, the modern discipline has too often adopted and reproduced those binaries and made little attempt to engage with feminist understandings of space.

In *Writing Women and Space: Colonial and Postcolonial Geographies*, a volume of essays edited by Alison Blunt and Gillian Rose (1994), feminist geographers use the experiences of female travellers such as Eliza Fraser and Mary Kingsley to theorize and explore women's role in the production of geographical knowledges (Mills, 1999). At once 'inside' and 'outside' the prevailing pattern of colonial discourse – thereby occupying a highly ambivalent position in relation to European colonization (Blunt and Rose, 1994; cf. Kearns, 1997) – these women were able to expose and subvert, to greater and lesser degrees, the representational economy of empire. Cheryl McEwan (2000) has similarly examined how the position of white women travellers shifted as they moved between home and abroad, exploring as they went their own sense of themselves as observers and agents of empire (Legg 2010).

Above all the testimony of these travellers highlights women's active agency – their status as 'subjects of history' – and adds much to recent efforts across the humanities and social sciences to de-colonize knowledge and 'revindicate' (Pratt, 1992: 2) the experiences and life-ways of the oppressed. It also reminds us that gender overlaps and intersects with other systems of domination, such as race and class, and that 'many of the world's women are oppressed by much more than patriarchy' (Sharp, 2009: 116). Research on 'moral masculinities' (Duncan, 2000; Lambert, 2005) and sexualized geographies (Phillips 1999; Nally 2004) are just two attempts to show how gendered modes of power intersect with other vectors of privilege and domination. Discussions of embodiment, abstraction, intersubjectivity and counter-hegemonic identities loom large in this research, illustrating the enormous influence of feminist criticism on colonial and postcolonial geographies.

Feminist geographers have also explored how space is fundamental to any exercise of power. The organization of space into 'public' and 'private' spheres, for instance, rests on the historically reinforced (and now taken for granted) associations between the domestic and the feminine (Duncan, 1996b; McDowell, 1999; Beckingham, 2012). Philip Howell's (2009a) research on the regulation of sex-workers in the university town of Cambridge, England is a case in point. Females walking public streets alone at night-time were presumed sexual miscreants and were frequently seized and detained by the local police and college authorities. Male students, on the other hand, even when caught *in flagrante delicto*, were commonly warned and released. Through the 'proctorial' system of policing, public space was coded and organized according to a very simple calculus: the innocence and sexual health of the university's male undergraduates was to be protected, while prostitute women – depicted as fallen, depraved and disorderly – were subject to surveillance, policing, medical inspection and incarceration (Howell, 2009: 17–18). This double standard of treating male undergraduates leniently and female sex-workers severely was captured in a revealing aphorism by *The Times*: 'His frailty is corrected by her liability' (cited in Howell, 2009a: 126). As Howell (2009a: 124–125) concludes:

> The emphasis on the 'public' evils of prostitution, the attack on the overt sexualisation of public space that soliciting entailed, is clearly evident in the equation of prostitution with streetwalking ... The Prostitute was portrayed here as the aggressor, the temptress, the instigator; and it was her public presence that was thus felt to be the real problem of prostitution.

Mona Domosh has similarly shown how the public promenades and walkways of nineteenth-century New York were highly scripted sites where gendered 'social mores' were both enacted and resisted. Bourgeois women, in particular, were threatened with losing their respectability by travelling on the streets of Broadway late at night (Domosh, 1998; see also Scobey, 1992). Domosh illustrates how women refused and challenged the disciplinary dictates of 'polite society' through minor acts of resistance, though the more interesting and salient point is that those norms were so deep rooted and entrenched that 'resistance' could only ever take the form of minor infractions and 'slight, everyday transgressions' (Domosh, 1998: 210). Today women routinely face ostracism, shame, intimidation, and physical violence in attempting to confront, in a more open fashion, the patriarchal values that structure identity behaviour in public space. Katherine Browne's (2004) important work on public toilets shows how 'gender ambiguous bodies' can be humiliated and publicly threatened for having supposedly entered the 'wrong space' (Creswell, 1996). Drawing on Judith Butler (1999), among others, Browne's work underscores the performative nature of gender and highlights the 'discontinutities between the sex with which an individual identifies and how others, in a variety of spaces, read their sex' (cf. Brown, 2000). The 'discontinuities' highlighted by Browne can sometimes be life threatening. Andrew Tucker's (2009) work on Cape Town illustrates how certain kinds of 'queer visibility' bring with them the risk of violent attack precisely because they are *seen* to challenge heterosexist white power that continues to structure social life in the post-apartheid South African state.

Gender and Dependency

Much of the work just discussed draws on feminist theory to reassess the practice of historical and cultural geography, including the historical reconstruction of Geography's own past. Clearly feminist scholars have taught historical geographers a great deal, but historical scholars might also add to contemporary discussions on gender and subjectivity. Haraway's point that we have failed to *historicize* and *relativize* the categories of gender and sex is an obvious call to appropriate history to think bodily identity anew. Not only should historical geographers join feminist scholars in exploring, say, the historical emergence of 'domestic' space (McClintock, 1995) or the splintering of the social world into rigidly controlled 'public' and 'private' spheres; they might also intervene to contest contemporary frameworks for asserting difference that rest upon, and are validated by, historical patterns of thought.

Geographers Emily Kaufman and Lise Nelson (2012) provide a wonderful example of this kind of scholarship. The authors trace the contemporary racialized and gendered image of the 'welfare queen' – popularized by US media – back to the influential English economist Thomas Robert Malthus. Malthus, as the authors explain, was no crackpot Victorian thinker. His views on welfare reform became the corner stone of the New English Poor Law and today he is regarded as one of the founders of modern economics. In his renowned *Essay on the Principle of Population* (1989: first published anonymously in 1798), Malthus attacked the existing method for delivering aid to the poor known as the Speenhamland system. Established by a group of Berkshire magistrates, but later applied nationwide, this early welfare system subsidized paupers on a scale linked to the current price of bread (cf. Polanyi, 2001). Malthus bitterly protested that the Speenhamland system all but guaranteed a basic income for the poor regardless of whether they worked. It was, he claimed, a 'gratuitous' allowance that only served to stimulate the growth of 'paupers' by allowing them to bear children without any personal responsibility for their maintenance.

Malthus offered precious little discussion of the economic and political realities that inform reproductive choices (Robbins, 1998). According to Kaufman and Nelson (2012), his arguments against public relief rest instead on a moral reading of 'population pressure', which depicts poor mothers as a threat to the collective health and welfare of the nation (cf. Davenport, 1995). By reproducing beyond their means the poor threatened corruption, decay, economic ruin and the eventual dislocation of conventional society. As Simon Commander (1986: 666) notes, the population principle 'was essentially a statement of human nature, and largely, a statement concerning fertility levels among the lower classes'.

Kaufman and Nelson (2012) trace this line of thinking through to late twentieth-century discussions of welfare reform in the United States. Focusing on public and media debates leading up to the passage of the Personal Responsibility and Work Opportunity Reconciliation Act of 1996, the authors show that 'national-level discourses depict the poor in pathologizing terms, as both repulsive and engaged in an expansion that must be checked in order to preserve or restore social order' (Kaufman and Nelson,

2012: 436). Charles Murray, a spokesperson for the American Enterprise Institute for Public Policy Research, is cited (Kaufman and Nelson, 2012: 437) as only one example: 'Reducing illegitimacy is not one of many desirable things to do. It is the prerequisite for rebuilding civic life in low-income black America, and for preventing a slide into social chaos in low-income white America.' Alarmist rhetoric of this kind formed part of a legitimizing apparatus that helped secure educational amendments promoting sexual abstinence, family caps on welfare payments to single mothers, and fiscal rewards to states that reduce their 'out-of-wedlock births'. This gendered 'ontology of scarcity and terror' (Kaufman and Nelson, 2012: 443) is not unique to welfare reforms passed under the Personal Responsibility and Work Opportunity Reconciliation Act of 1996. The move to 'pathologize the poor' is also evident in Mitt Romney's much debated division of US society into 'makers' and 'takers', while in the UK recent spending cuts to welfare programmes followed spurious efforts to bifurcate public life into groups of 'strivers' and 'skivers' (Williams, Z., 2013).

The Victorian distinction between the 'deserving' and 'undeserving' poor echoes strongly in contemporary discussions of 'dependency culture' and allied attempts to dismantle the remaining vestiges of the welfare state. The gendering of this debate in the twenty-first century is no less explicit than it was in Malthus's time: anecdotes of a hidden 'scrounger nation' jostle with sneering remarks about 'broken homes' and the need to hold parents (read 'mothers') accountable for the social failings of their children (consider, for example, the public discourse of blame following the riots in England during the summer of 2011, when 'feral youths' and 'failing families' were quickly cited as the 'real cause' of the conflagration). This repeated positioning of working-class mothers as the locus of national concerns around morality is open to critical modes of historical reflection.

Conclusion

As this chapter was being drafted, a news item announced that Virgin Media in the US has withdrawn an advertisement that seemingly makes light of sexual assault. The offensive advert depicts a man hugging a gift in one arm whilst holding shut the eyes of a woman with his other hand. The accompanying caption declaims: 'The gift of Christmas surprise. Necklace? Or chloroform?' Richard Branson, the business tycoon who owns Virgin media, issued a terse public apology, declaring the commercial 'ill-judged' and a 'dreadful mistake' (Williams, M., 2013). This news item also surfaces amidst a week of furious protests over the vicious gang rape and murder of Jyoti Singh Pandey, a 23 year-old medical student from Delhi. The protests spread beyond India to Nepal, Sri Lanka, Pakistan and Bangladesh, sparking a frenetic debate on public attitudes to rape and sexual violence. The news reminds us that there is a culture that colludes with sex and gender-based violence (Dustin, 2013). Arguably the thin end of the wedge is the repression of the social construction of bodily identity – a move that excludes non-normative forms of subjectivity and paves the way for toxic modes of identification including, but not limited to, androcentrism, patriarchy, sexism, and misogyny. The social cost of 'mistaking cultural forces for natural ones' – to cite Braun and Moeckli's (2001) felicitous phrase – is truly enormous. In thinking the present backwards historians and historical geographers are well placed to explain how 'natural orders' are socially produced and how a relativized understanding of our own subjectivity is the first step toward building a more inclusive, less violent, and more just society.

KEY POINTS

- In geography feminist scholars have shown how sex–gender categories are historically and socially constructed. For them biology is not destiny and one's bodily identity is never a natural fact.
- Feminist geographers have challenged the traditional idea of geography as a male enterprise. Much of this research highlights women's active agency in the production of historical geographic knowledge.
- From a range of different traditions, geographers have shown how the organization of space into 'public' and 'private' spheres rests on taken for granted associations between the domestic and the feminine.
- Geographers have also begun to explore how contemporary framings of gender difference rest upon, and are validated by, historical patterns of thought.

References

Beckingham, D. (2012) 'Gender, space and drunkenness: Liverpool's licensed premises, 1860–1914', *Annals of the Association of American Geographers*, 102(3): 647–666.

Blunt, A. and Rose, G. (eds) (1994) *Writing Women and Space: Colonial and Postcolonial Geographies*. London: Guilford.

Bondi, L. and Davidson, J. (2003) 'Troubling the place of gender', in K. Anderson, M. Domosh, S. Pile and N. Thrift (eds) *Handbook of Cultural Geography*. London: Sage, pp. 325–343.

Brown, M. (2000) *Closet Space: Geographies of Metaphor from the Body to the Globe*. London: Routledge.

Browne, K. (2004) 'Genderism and the bathroom problem: (re)materialising sexed sites, (re)creating sexed bodies', *Gender, Place and Culture*, 11(3): 331–346.

Butler, J. (1986) 'Sex and gender in Simone de Beauvoir's *Second Sex*', *Yale French Studies*, 72: 35–49.

Butler, J. (1999) *Gender Trouble: Feminism and the Subversion of Identity*. New York: Routledge.

Commander, S. (1986) 'Malthus and the theory of "unequal powers": population and food production in India 1800–1947', *Modern Asian Studies*, 204(4): 661–701.

Cresswell, T. (1996) *In Place/Out of Place*. Minneapolis: University of Minnesota Press.

Davenport, R. (1995) 'Thomas Malthus and maternal bodies politic: gender, race, and empire', *Women's History Review*, 4(4): 415–439.

De Beauvoir, S. (1989) *The Second Sex* (Introduction by Deirdre Bair and translated by H. M Parshley). New York: Vintage Books.

Domosh, M. (1998) 'Those "gorgeous Incongruities": polite politics and public space on the streets of nineteenth-century New York', *Annals of the Association of American Geographers*, 88(2): 209–226.

Domosh M. (1991) 'Toward a feminist historiography of geography', *Transactions of the Institute of British Geographers*,16 (1): 95–104.

du Plessix Gray, F. (2010) 'Dispatches From the Other', *New York Times*, 27 May. http://www.nytimes.com/2010/05/30/books/review/Gray-t.html?pagewanted=all&_r=0 (Accessed 11 January 2013).

Duncan, J. (2000) 'The struggle to be temperate: climate and "moral masculinity" in mid-nineteenth-century Ceylon', *Singapore Journal of Tropical Geography*, 21(1): 34–47.

Duncan, J.S. (2007) *In the Shadows of the Tropics: Climate, Race and Biopower in Nineteenth Century Ceylon*. Aldershot: Ashgate.

Duncan, N. (1996a) *BodySpace: Destabilising Geographies of Gender and Sexuality*. London: Routledge.

Duncan, N. (1996b) 'Renegotiating gender and sexuality in public and private space', in N. Duncan, *BodySpace: Destabilising Geographies of Gender and Sexuality*. London: Routledge, pp. 145–154.

Dustin, H. (2013) 'Jimmy Savile report: this must mark a turning point', *Guardian*, 11 January. www.guardian.co.uk/commentisfree/2013/jan/11/jimmy-savile-report-turning-point (accessed 11 January 2013).

Gibson-Graham, J.K. (1994) 'Stuffed if I know!': reflections on post-modern feminist social research', *Gender, Place and Culture*, 1(2): 205–224.

Haraway, D. (1991) *Simians, Cyborgs and Women: The Reinvention of Nature*. London: Free Association Books.

Haraway, D. (1992) 'The promises of monsters: a regenerative politics for inappropriate/d others', in L. Grossberg, C. Nelson and P.A. Treichler (eds) *Cultural Studies*. New York: Routledge, pp. 295–337.

Howell, P. (2009a) *Geographies of Regulation: Policing Prostitution in Nineteenth-Century Britain and the Empire*. Cambridge: Cambridge University Press.

Howell, P. (2009) 'Sexuality', in R. Kitchin and N. Thrift (eds) *International Encyclopedia of Human Geography, Volume 10*. Oxford: Elsevier, pp. 119–124.

Kaufman, E. and Nelson, L. (2012) 'Malthus, gender and the demarcation of "dangerous" bodies in 1996 US Welfare Reform', *Gender, Place and Culture*, 19(4): 429–448.

Kearns, G. (1997) 'The imperial subject: geography and travel writing in the work of Halford Mackinder and Mary Kingsley', *Transactions of the Institute of British Geographers*, 22: 450–472.

Kobayashi, A. (2001) 'The nature of race', in N. Castree and B. Braun (eds) *Social Nature: Theory, Practice and Politics*. London: Blackwell, pp. 64–83.

Kuruvilla, C. (2012) 'Pope Benedict denounces gay marriage during his annual Christmas message', *New York Daily News*, 22 December. www.nydailynews.com/news/world/pope-denounces-gay-marriage-annual-xmas-message-article-1.1225960#ixzz2GjSPokCO (accessed 22 December 2012).

Lambert, D. (2005) *White Creole Culture, Politics and Identity during the Age of Abolition*. Cambridge: Cambridge University Press.

Legg, S. (2010) 'An intimate and imperial feminism: Meliscent Shephard and the regulation of prostitution in colonial India', *Environment and Planning D: Society and Space*, 28(1): 68–94.

Malthus, T.R. (1989) *An Essay on the Principle of Population Volume II* (ed. Patricia James). Cambridge: Cambridge University Press.

Márquez, G.G. (1988) *Love in the Time of Cholera*. New York: Alfred A. Knopf.

McClintock, A. (1995) *Imperial Leather: Race, Gender and Sexuality in the Colonial Context*. London: Routledge.

McDowell, L. (1999) *Gender, Identity and Place*. Cambridge: Polity.

McEwan, C. (2000) *Gender, Geography and Empire: Victorian Women Travellers in West Africa*. London: Ashgate.

Mill, J. S. (1869) *The Subjection of Women*. New York: D. Appleton Co.

Mills, S. (1999) 'Gender and colonial space', *Gender, Place and Culture*, 3(2): 125–147.

Mitchell, D. (2000) *Cultural Geography: A Critical Introduction*. Oxford: Wiley-Blackwell.

Moeckli, J. and Braun, B. (2001) 'Gendered natures: feminism, politics and social nature,' in N. Castree and B. Braun, *Social Nature: Theory, Practice and Politics*. London: Blackwell, pp. 112–133.

Nally, D. (2004) 'Incorrigible Venice and the war against cliché', *Environment and Planning D: Society and Space*, 22(1): 295–312.

Phillips, R. (1999) 'Writing travel and mapping sexuality: Richard Burton's sotadic zone', in J. Duncan and D. Gregory (eds) *Writes of Passage: Reading Travel Writing*. London: Routledge, pp. 70–91.

Polanyi, K. (2001) *The Great Transformation: The Political and Economic Origins of Our Times*. Boston: Beacon Press.

Pope Benedict XVI (2012) 'Address of His Holiness Benedict XVI on the occasion of Christmas greetings to the Roman Curia', Clementine Hall, 21 December. www.vatican.va/holy_father/benedict_xvi/speeches/2012/december/documents/hf_ben-xvi_spe_20121221_auguri-curia_en.html (accessed 21 December 2012).

Pratt, M.L. (1992) *Imperial Eyes: Travel Writing and Transculturation*. London: Routledge.

Robbins, P. (1998) 'Population and pedagogy: the geography classroom after Malthus', *Journal of Geography*, 97(6): 241–252.

Rose, G. (1993) *Feminism and Geography*. Cambridge: Polity.
Scobey, D. (1992) 'Anatomy of the promenade: the politics of bourgeois sociability in nineteenth-century New York', *Social History*, 17(2): 203–227.
Sharp, J.P. (2009) *Geographies of Postcolonialism*. London: Sage.
Spivak, G. (1987) *In Other Worlds: Essays in Cultural Politics*. New York: Methuen.
Tucker, A. (2009) *Queer Visibilities: Space, Identity and Interaction in Cape Town*. Oxford: Blackwell.
Williams, M. (2013) 'Virgin mobile US takes down Christmas advert suggesting sexual assault', *Guardian*, 9 December.www.guardian.co.uk/media/2012/dec/09/virgin-mobile-us-holiday-ad (accessed 9 December 2012).
Williams, Z. (2013) 'Skivers v strivers: the argument that pollutes people's minds', *Guardian*, 9 January. www.guardian.co.uk/politics/2013/jan/09/skivers-v-strivers-argument-pollutes (accessed 9 January 2013).
Wollstonecraft, M. (1972) *A Vindication of the Rights of Woman with Strictures on Political and Moral Subjects* (Edited by Miriam Brody Kramnick). Harmondsworth: Penguin.

FURTHER READING

Blunt, A. and Rose, G. (eds) (1994) *Writing Women and Space: Colonial and Postcolonial Geographies*. London: Guilford.
Howell, P. (2009) *Geographies of Regulation: Policing Prostitution in Nineteenth-Century Britain and the Empire*. Cambridge, Cambridge University Press.
Kaufman, E. and Nelson, L. (2012) 'Malthus, gender and the demarcation of "dangerous" bodies in 1996 US Welfare Reform', *Gender, Place and Culture*, 19(4): 429–448.
Rose, G. (1993) *Feminism and Geography*. Cambridge: Polity.
Tucker, A. (2009). *Queer Visibilities: Space, Identity and Interaction in Cape Town*. Oxford: Blackwell.

Section 4 The Built Environment

10 NATURE AND THE ENVIRONMENT

Ulf Strohmayer

Introduction

For the longest time, historical geography did not include 'nature' or 'the environment' amongst the topics explicitly selected for closer scrutiny. In close analogy to the majority of historical studies, historical geography was all about 'man' and his, occasionally her, lasting imprint on the planet. In other words: if we accept the time-worn, polarized differentiation between 'nature' and 'culture' as a valid discrimination between two independent analytical domains (and, as we shall see, there are many good reasons for not accepting it), historical geography firmly sided with 'culture' in its pursuit of geographical knowledge. 'Nature' became a topic only where and if it had been formed according to human creativity and tradition (as can be seen throughout this book, the notion of 'form' is indeed central to the practices emanating forth from within historical geography). Thus it is that 'gardens', 'parks' or 'cultural landscapes' more generally emerged as legitimate subject matters for historical geographers, while 'woods' or 'bogs' did not. Where they did, as in George Perkins Marsh's foundational *Man and Nature* (1864), they mattered because either 'man' had rendered them obsolete through processes of deforestation, overgrazing or the like, or because, like certain species of wildlife or plants, they were introduced to particular places by human activity. Thus not surprisingly, reconstructing this influence through painstaking construction of 'before' and 'after' pictures became the task at hand.

For some time, such a manner of thinking made perfect sense: we must remember that 'geography' was principally defined as a discipline populating the nexus between 'environment' or 'nature' and humankind. If this nexus is perhaps best exemplified by the historically specific foundation of National Parks in the nineteeth century (see Nash 1967), it is surprising that for the longest time the key aspect of construction and management did not appear to be of interest to historical and other geographers; or indeed to those historians writing what is called 'environmental history' – a close cousin to the attempts made within the discipline proper.

For human geographers, and historical geographers among them, this nexus was thus conceptualized primarily in the form of 'human impacts', where the absence of human design or formatting relegated academic interests to a realm other than the one prescribed by historical geography (Williams, 1994). Thus 'environmental history' and again 'natural history' were both different from 'historical geography', also because the time frames attaching to their respective subject areas were much longer than anything human geographers were willing to engage with.

Key to this approach was the notion of 'landscape' that historical geographers, not unlike their colleagues in cultural geography, embedded centrally within their works. 'Landscapes', not 'nature' or 'environment', became the embodiment of the nexus between humankind and the respective contexts within which historically relevant facts and processes unfolded. As Baker (1992) has argued conclusively, the transformation of 'natural' into 'cultural' landscapes became a central occupation for those historical geographers occupied with large-scale transformative processes. By definition, 'nature' thus became 'transformed nature', or what Smith, following the eighteenth-century French naturalist (and founder of biogeography) Buffon, has called 'second' nature (Smith, 1984: 46–55). Baker (2003: 78) also alludes to this crucial distinction when he insists on the difference between 'landscape' and 'environment'.

Landscape-as-Such

In the majority of cases, traditional concerns within historical geography attached themselves to the genesis of particular landscapes, be they urban or rural in character. As such, historical geographers treated 'nature' as the mouldable material that yielded different results depending on how, and using what kind of tools, humankind 'worked' on it. Crucially, the result of such labour was thought to be accessible to historically trained geographers because it was visible in the particular landscapes thus created. Understanding how such landscapes came to *look* the way they presented themselves to, and were represented for, both lay and academic onlookers was the key objective of historical geography (Lewis, 1983). Hence the fascination that visual sources of whatever kind – from maps to paintings to photographs – have and continue to exert on those practising the trade (see Chapter 23). The undeniable aesthetic dimension that came to the fore in the ensuing incorporation of 'nature' – albeit often merely as a background phenomenon – into a body of work associated with historical geography often led to the production of insightful but descriptive scholarship. At the same time, it often led to a reification of associations with 'nature' that were culturally produced. In this process, natural and cultural landscapes became deeply and intrinsically embedded within grand histories that expressed, supported and justified nationalized narratives of conquest and belonging. The centrality accorded to wooded lands and forests in the narratives of the German nation, as Simon Schama has demonstrated by way of example (Schama, 1996; see also Bassin, 2005), matches the notion of 'wide open spaces' in the construction of the American West (Hoelscher, 1999; Harris, 2004). Similarly, geographers have paid close attention to associated practices of naming landscapes by restoring or inventing designations or place markers (Azaryahu, 1997; Azaryahu and Golan, 2001; see also Holdsworth, 2002 and the chapters in Section 5).

Landscape-as-Process

The move away from a largely implied to a critical reading of 'nature' in the work of historical geographers was closely associated with developments across the human sciences during the course of the last three decades. As happened in many other fields of human geographic inquiry, the so-called 'cultural turn' left a lasting imprint on the manner in which geographers saw fit to analyse historically shaped, as well as contemporary landscapes. And while it is not always easy or indeed fruitful to seek to determine whether any particular piece of writing should count as a contribution to the field of 'cultural' or 'historical' geography, increasingly notions of culture or culturally inspired change entered into the conceptual worlds invoked in historically motivated attempts to make sense of the world. Crucially, 'culture' was no longer conceptualized as a 'positive' deposition or depository of human expression – a museum of sorts – but as an actively shaped and contributing factor that helped explain how and why people acted the way they did (Demeritt, 1994). Initially at least, and inspired in part by the publication of Cosgrove's seminal work on the symbolic importance of historical landscapes (1984), this shift prompted practitioners working in the sub-field to focus on less materially rooted re-presentations in history as their main source material. Over time, such initial restrictions have given way to the current, rather broad and eclectic engagements with nature in the writings of historical geographers. What unites this body of work is a focus on the historically rooted and contextualized *production* of nature.

Given this emphasis on constructed change, there is little wonder that a significant number of publications have appeared in print, which interpret the city as a crucial site in the historical formation of natures. Not only, as Chapter 12 will argue further on, did the city become an object that was perfectly suited to epistemic desires imbued with the 'cultural turn' (with 'urbanization' processes and 'urban change' becoming a chosen and preferred field of study; Jenkins, 2006). As concerns our present focus, the city furthermore emerged as what Holdsworth metaphorically christened an 'octopus' reaching into environments beyond the city in search for resources (2004). A key target of research became the time-honoured contrast between 'nature' (broadly construed) and urban forms of life, which was increasingly thematized in terms of a symbiotic and historically contingent constellation. One of the best – certainly one of the most influential – studies to break the accepted mould was William Cronon's *Nature's Metropolis* (1991), a magisterial study of the 'historicization' of nature in the development of Chicago (Cronon, 1991). Other works to re-cast the historical development of urban environments with reference to particular conceptualizations of nature include Nicholas Green's *The Spectacle of Nature* (1990) and Matthew Gandy's *Concrete and Clay* (2002). Especially urban parks resurfaced as spaces exemplifying the nexus between 'nature' and 'culture', now increasingly recast as socially produced spaces that involved labour, rather than aesthetic principles (Debarbieux, 1998).

Nature as Product

Beyond the confines of a narrowly conceived 'cultural turn', geographers began to conceptualize 'nature' as being more than the residual category left over after the work of culture and history had run its analytical course. This 'more than' took on the form of some kind of 'production' or another; thus 'nature' emerged as 'socially' or 'culturally' produced (Fitzsimmons, 1989; Castree, 1995; Kong and Yeoh, 1996; Demeritt, 2001). In the context of such debates, it is no accident that historical geographers increasingly turned their attention especially to the way in which historical constructions and uses of nature help to understand key processes that shape historical and contemporary geographies. Anderson's (2003) paper on the mutually supporting links between white colonial identities, their 'civilizing' mission and the construction of nature within a historical setting is thus indicative of a larger trend in historical geography that transcends the narrow focus of the present chapter: crucial for these and related explorations at the nexus of geography and history is their desire to employ historical evidence to shed light on broader, non-nationalized processes such as colonialism, imperialism, modernity or the like, all of which feature prominently in the present volume (see also Driver and Martins, 2003 for a collection of essays exemplifying this particular interpretation of historical geography). Part and parcel of many of these narratives is an overt emphasis on contestation as a motivating element in the construction of historically evolving landscapes. Increasingly seen to be embedded in political processes, 'historical landscapes' lost much of the natural sheen that attached to them in the past; in its stead, historical geographers focused on the eminently political production of landscapes within the context of broader contested processes such as 'modernization', 'capitalism' or 'globalization', to name but a few key processes. This, too, is not an accident: deprived of a 'nationalist' context that served, *inter alia*, to streamline and naturalize any resulting narratives, 'nature' and the 'environment' are much more likely to emerge as concepts that point towards contested geographies and are thus embedded no less firmly in the historical hierarchies that formed the centre of our attention in the preceding section of this book.

The recognition of 'contestation' as an important factor contributing to the shaping of 'nature' – in historical geography and beyond – has the intended consequence of de-naturalizing a number of commonly accepted ways of seeing and understanding 'nature'. Thus cut loose from traditional ways of contextualizing historical geographies of nature, these latter were now increasingly contributing to different narratives. 'Gender' is one such exemplary narratival context: spurred by the publication of Merchant's *Death of Nature* (1980), conceptually conscious investigations into historical constructions of nature began to emerge (see Nash, 2000) and shape the way geographers and other spatially minded scholars thought about nature. A similar historiography of the ways in which geographers approach history shapes many chapters in this volume: the notion of 'conceptually informed' knowledge would be an apt description of such endeavours.

Representing Nature

Just as crucially, the notion of 'contested' natures and environments also attaches to representations of natural worlds; in a historically motivated context, where access to evidence is often mediated by the quality and quantity of archival presences, representations of both nature and of 'natural' environments are often impossible to separate from what scholars assumed to be thus represented (see especially Schein, 1993; Raivo, 1997; Withers, 2000; Hagen, 2004). Conversely, a representation of nature can lead us towards more richly nuanced histories than the ones suggested or available in their absence.

Take, by way of closing example, the representation of nature on a postcard (see Figure 10.1). In one sense, 'nature' is readily available here in the form of trees, shrubs and terrain. In addition, we see a bridge, a number of people and some kind of temple in the central distance. We are also informed by the title of the postcard that what we see was located in the city of Paris, that the bridge is a suspension bridge and that the space is referred to as the 'Buttes Chaumont', allowing us to infer (assuming we can read French and are able to 'place' the geographical designation 'Paris') that we are indeed looking at a park in an urban environment. Furthermore, those acquainted with fashion may well be able to decode from the clothing worn by the three visible persons that we are looking at a representation of nature dating around the nineteenth-century fin-de siècle.

Figure 10.1 Postcard Paris: Buttes Chaumont – Le pont suspendu, ca. 1905

At the same time, however, a representation can only show so much – and often, in showing something in a particular manner, effectively blanks out other things. This is perhaps most obvious in the form of the frame and the angle adopted for and by a particular representation: the absence of houses in the postcard effectively eliminates the urban context of the scene depicted. It is no less 'at work' in the non-existence of maintenance work from the picture (see Strohmayer, 2006), in the 'blending' of technology and culture into nature (note the organic link between wood and supportive iron material on the bridge or the placing of an ostensibly 'Greek' temple on top of the mountain), and in the gendered depiction of those people using the bridge, with two male figures purposely walking though nature while the female is accorded the time leisurely to enjoy it. The fact that of these people, it is only the female figure that was afforded a post-production colouring treatment furthermore adds to the 'leisurely' feel attaching to her pose (itself invisible in a black-and-white representation) (see also Chapter 9).

Lastly, what such representations fail to acknowledge, show or allude to is the *production of nature* – in the case of the Parc des Buttes-Chaumont: the literal design of a naturally appearing former productive environment in the form of a quarry yielding a large part of the construction material required for the Haussmannian transformation of Paris during the 1850s and 1860s. 'Nature' here contributes to a host of nominally external but functional linked contexts such as the re-imagination of France as an imperialist and modern society, of Paris as an organic, conflict-free city populated by non-rebellious, harmoniously co-existing inhabitants. Historical geographers, as discussed in more detail in Section 8, need thus be wary of forms of representation in general – and of represented forms of nature more particularly: while they provide unique perspectives and insights into past environments, they need to be accorded a critical sense of scrutiny and contextual awareness to be useful pointers in the analysis of past geographies.

KEY POINTS

- In recent years, the interest shown by historical geographers towards the natural world has shifted from a largely aesthetic or functional focus towards one concerned with notions of contestation and construction.
- Rather than seeing in 'nature' or the 'environment' a mere background expression, often of nationalized tendencies, both concepts are seen today to offer key insights into the historical genesis of social, cultural and economic realities.
- As part of such processes, 'nature' emerges as a thoroughly constructed and often contested category.

References

Anderson, K. (2003) 'White natures: Sydney's Royal Agricultural Show in post-humanist perspective', *Transactions of the Institute of British Geographers* NS28, 422–441.

Azaryahu, M. (1997) 'German reunification and the politics of street names: the case of East Berlin', *Political Geography*, 16(6): 479–493.

Azaryahu, M. and Golan, A. (2001) '(Re)naming the landscape: the formation of the Hebrew man of Israel, 1949–1960', *Journal of Historical Geography*, 27(1): 178–195.

Baker, A. (1992) 'Introduction: on ideology and landscape' in A.R.H. Baker and G. Biger (eds) *Ideology and Landscape in Historical Perspective*. Cambridge: Cambridge University Press.

Baker, A. (2003) *Geography and History: Bridging the Divide*. Cambridge: Cambridge University Press.

Bassin, M. (2005) 'Blood or soil? The volkisch movement, the Nazis, and the legacy of Geopolitik', in F.-J. Brüggemeier, M. Cioc and T. Zeller (eds) *How Green Were*

the Nazis? Nature, Environment, and Nation in the Third Reich. Athens, OH: Ohio University Press, pp. 204–242.
Castree, N. (1995) 'The nature of produced nature', *Antipode*, 27: 12–48.
Cosgrove, D. (1984) *Social Formation and Symbolic Landscape*. London: Croom Helm.
Cronon, W. (1991) *Nature's Metropolis: Chicago and the Great West*. New York: Norton.
Debarbieux, B. (1998) 'The mountain in the city: social uses and transformations of a natural landform in urban space', *Ecumene*, 5(4): 399–431.
Demeritt, D. (1994) 'Ecology, objectivity and critique in writings on nature and human societies', *Journal of Historical Geography*, 20: 22–37.
Demeritt, D. (2001) 'Being constructive about nature,' in N. Castree and B. Braun (eds) *Social Nature. Theory, Practice and Politics*. Oxford: Blackwell, pp. 22–40.
Driver, F. and Martins, L. (eds) (2003), *Tropical Visions in an Age of Empire*. Chicago: Chicago University Press.
Fitzsimmons, M. (1989) 'The matter of nature', *Antipode*, 21: 106–120.
Gandy, M. (2002) *Concrete and Clay: Reworking Nature in New York City*. Cambridge, MA: MIT Press.
Green, N. (1990) *The Spectacle of Nature: Landscape and Bourgeois Culture in Nineteenth-century France*. Manchester: University of Manchester Press.
Hagen, J. (2004) 'The most German of towns: creating an ideal Nazi community in Rothenburg ob der Tauber', *Annals of the Association of American Geographers*, 94(1): 207–227.
Harris, C. (2004) 'How did colonialism dispossess? Comments from an edge of Empire', *Annals of the Association of American Geographers*, 94(1): 165–182.
Hoelscher, S. (1999) 'From sedition to patriotism: performance, place and the reinterpretation of American ethnic identity', *Journal of Historical Geography*, 25(4): 534–558.
Holdsworth, D. (2002) 'Historical geography: the ancients and the moderns – generational vitality', *Progress in Human Geography*, 26: 671–678.
Holdsworth, D. (2004) 'Historical geography: the octopus in the gardens and in the fields', *Progress in Human Geography*, 28: 528–535.
Jenkins, L. (2006) 'Utopianism and urban change in Perreymond's plans for the rebuilding of Paris', *Journal of Historical Geography*, 32: 336–351.
Kong, L. and Yeoh, B. (1996) 'Social construction of nature in urban Singapore', *Southeastern Asian Studies*, 34: 402–423.
Lewis, P. (1983) 'Learning from looking: geographic and other writing about the American cultural landscape', *American Quarterly*, 35: 242–261.
Marsh, G.P. (1864) *Man and Nature, or, Physical Geography as Modified by Human Action*. New York: Charles Scribner.
Merchant, C. (1980) *The Death of Nature: Women, Ecology, and the Scientific Revolution*. London: Harper & Row.
Nash, C. (2000) 'Environmental history, philosophy and difference', *Journal of Historical Geography*, 26(1): 23–27.
Nash, R. (1967) *Wilderness and the American Mind*. Yale, CT: Yale University Press.
Raivo, P.J. (1997) 'The limits of tolerance: the Orthodox milieu as an element in the Finnish cultural landscape, 1917–1939', *Journal of Historical Geography*, 23(3): 327–339.

Schama, S. (1996) *Landscape and Memory*. New York: Vintage.
Schein, R. (1993) 'Representing urban America: nineteenth century views of landscape, space, and power', *Environment and Planning D: Society and Space*, 11(1): 7–21.
Smith, N. (1984) *Uneven Development*. Oxford: Blackwell.
Strohmayer, U. (2006) 'Urban design and civic spaces: nature at the Parc des Buttes-Chaumont in Paris', *Cultural Geographies*, 13(4): 557–576.
Williams, M. (1994) 'The relations of environmental history and historical geography', *Journal of Historical Geography*, 20: 3–21.
Withers, C. (2000) 'Authorising landscape: "authority", naming and the Ordnance Survey's mapping of the Scottish Highlands in the nineteenth century', *Journal of Historical Geography*, 26(4): 532–554.

FURTHER READING

Debarbieux, B. (1998) 'The mountain in the city: social uses and transformations of a natural landform in urban space', *Ecumene*, 5(4): 399–431.
Gandy, M. (2002), *Concrete and Clay: Reworking Nature in New York City*. Cambridge, MA: MIT Press.
Green, N. (1990) *The Spectacle of Nature: Landscape and Bourgeois Culture in Nineteenth-Century France*. Manchester: University of Manchester Press.
Swyngedouw, E. (1999), 'Modernity and hybridity: nature, regeneracionismo, and the production of the Spanish waterscape, 1890–1930', *Annals of the Association of American Geographers*, 89: 443–465.

11 MAKING SENSE OF URBAN SETTLEMENT

Yvonne Whelan

Introduction

While today historical geography is a diverse field of human geography, as evidenced by the varied range of conceptual material considered in this book, it is much more than a discrete thematic strand. On the contrary, historical geography is economic, cultural, social and political, urban and rural, and it touches on a whole range of concerns. This is especially evident when it comes to questions relating to human settlement, population, cities and urbanization. Such topics have provided a rich seam of research for geographers working across a range of both urban and rural settings and over many different time periods. For some, the focus has rested on understanding historical settlement patterns of land use and on charting phases of growth, development and decline. Many rural historical geographers, for example, have analysed settlement patterns, drawing attention to phases of continuity and change in a variety of environmental, demographic and cultural contexts. For others, the urban domain has been the chief focus of inquiry and given that the city is one of the most powerful cultural productions of civilization, it is perhaps not surprising that geographers have been at the forefront of attempts to explore the complexity of urban spaces. A whole host of questions have spurred on research in this field, as geographers have attempted to account for how and why urban settlements are spatially organized in the ways that they are. This chapter reflects further on the processes and patterns that characterize urbanization, as well as on some of the trajectories that research has taken when it comes to making sense of one of the chief by-products of urbanization, namely the city and the townscape.

Geographies of Urban Settlement

In June 2005 the Campaign to Protect Rural England (CPRE) published a report entitled *Your Countryside, Your Choice* in which they envisaged a landscape with no distinction between town and country. Instead the report presented an anodyne vision of a landscape comprising out-of-town retail parks, 'meandering housing estates, ring roads, the backs of gardens, streetlights, signs and masts. By accident rather than design, much of England has become an anywhere place, unloved and unloving – an homogenous exurbia, in which everywhere looks the same as everywhere else' (Kingsnorth, 2005: 3). The report observed that

> the landscape is spattered and blotched with housing and sheds of all colours but mostly large, while what remains of open land is riddled with fairways, paddocks and shimmering polythene. The varicose network of roads pervades all, ceaselessly coursing with traffic from fat grey arteries to the writhing filigree of the cul-de-sac. (Kingsnorth, 2005: 2)

Over the course of the second half of the twentieth century our world has been transformed into a predominantly urban domain, characterized by an overwhelmingly urban population residing in giant cities. The stealthy urbanization referred to in the CPRE report is radically transforming the landscapes in which we live and bringing with it major implications for rural spaces as a total of 21 square miles of countryside is lost to development every year. This 'creeping urbanization' is not peculiar to Britain of course where over 80 per cent of the population now live in urban areas, but it is also reflected more widely across the globe. Presently, the urban population stands at just over 70 per cent, while the UN projects that some 60 per cent of the world's population will be living in cities by the year 2030. So, it seems certain that increased urbanization will be one of the key defining characteristics of life in the twenty-first century.

Let us now contrast this contemporary situation with that of Western Europe in the sixteenth century. Then, only six European cities had populations of greater than 100,000 with the vast majority living in small towns, villages and rural settings. The rise of capitalism in the sixteenth and seventeenth centuries, together with the onset of a concerted era of nation-building, ensured that the settlement system of Western Europe began to change. Gradually, what was once 'an area of mainly small cities with market areas limited by access from their agricultural hinterlands [...] became a region of rapid city growth within powerful nation states' (Lawton, 1989: 2). Despite this rapid growth, however, by the dawn of the nineteenth century most of Britain's population remained largely rural. In 1801, for example, the population of England and Wales was less than 9 million, with just 17 per cent living in the 15 towns of over 20,000 inhabitants. Only London, with it population of around 865,000 had more than 100,000 inhabitants (Daniel and Hopkinson, 1989: 66). It was not until the industrial revolution and the concomitant rise of large manufacturing bases across Europe that population began to grow rapidly and Western Europe's lead in the new urbanization was confirmed (Lawton, 1989: 3). As Badcock (2002: 21–22) observes:

> Urbanisation in Britain led the way. In the space of 90 years, between 1801 and 1891, the proportion of the English population living in towns rose from one-quarter to three-quarters – up from 17 per cent in 1801 to 61 per cent by 1911 for towns over 20,000. Before 1800 London was the only city over 100,000 but by 1911, 44 other British cities shared that same status.

During the nineteenth century, therefore, population rose, especially between 1840 and 1880 as a consequence of, and in response to, equally rapid industrialization. By 1891, Britain's population had reached 29 million, and the population of London stood at 6.5 million, with Birmingham and Manchester increasing ten-fold to 760,000 and 645,000 respectively by 1901 (Daniel and Hopkinson, 1989: 66). Such urbanization, the term used to describe an increase in the proportion of people living in urban areas compared to rural areas, 'was one of the hallmarks of modernization and came to be one of the symbols of economic and social change during the industrial revolution' (Lawton and Pooley, 1992: 90). It was also the product of a whole series of competing forces, not least the fact that the high death rates were exceeded by even higher birth rates leading to a natural increase, but also because of high levels of rural–urban migration. The transformation of small-scale cottage industries into much larger-scale factories ensured that towns and cities that were located in areas with good natural resources prospered and expanded, serving in turn as a 'pull factor' for an increasingly landless peasantry. Concurrently, a wide range of agrarian improvements and increased mechanization led to rising levels of rural unemployment, creating a series of 'push factors' that forced more and more people to move from the countryside to new urban areas in search of work (Daniel and Hopkinson, 1989: 66). It is also important to bear in mind, however, that the urbanization of Britain in the nineteenth century was part of a much broader and interdependent global process. As Badcock (2002: 22) points out,

> urbanisation and city-building in Europe, on the one hand, and in far-flung colonies, on the other, had as much to do with the asymmetrical flows of labour, capital and commodities between the core metropolitan powers and the colonies as between the countryside and the growing cities within the domestic economy.

Equally important to consider is the fact that urbanization occurred at very different rates throughout the world and was spurred on by quite different sets of regional circumstances. So, although industrialization was the chief spur to urbanization in Britain, in the New World it had its beginnings in mercantilism and manufacturing did not become the cornerstone of urban growth in the United States until well into the 1860s (Badcock, 2002: 22).

Modelling Urban Land Use

The processes of urbanization that I referred to above and the consequent rapid growth in cities in Western Europe and the US during the nineteenth and early twentieth centuries, prompted geographers, sociologists and scholars from allied disciplines, to consider different ways of making sense of cities. For one group of urban sociologists operating in Chicago in the early decades of the twentieth century the city became a particular focus of inquiry in their attempts to model internal organization and land use and to account for the social and cultural mosaic of the city (Johnston, 1971). Although now largely discredited, these early initiatives provide us with some important foundational attempts to explain the urban landscape by foregrounding how it acted as a container of different, sometimes segregated, regions of land use and socio-economic zones.

The Chicago School of Urban Ecology grew out of the United States' first Department of Sociology which was established at the University of Chicago in 1913 (Johnston, 1971; Bulmer, 1984). Spurred on by the burgeoning growth and industrialization of US cities in the 1920s, this heady period of sociological research was spearheaded by Robert E. Park and Ernest Burgess. Together with other researchers of the Chicago School, they variously probed the internal geography of the city, oftentimes utilizing painstaking methods of ethnographic research in order to analyse some of the social problems that went hand in hand with increased urbanization. The fact that 1920s Chicago was experiencing an ever-growing influx of migrants and rapid urbanization made it an ideal testing ground. Some Chicago School scholars advanced the idea of an urban ecology, likening society to an organism, and looked to natural ecosystems and Darwin's theory of evolution to provide an analogue of the processes producing natural areas in cities (Badcock, 2002: 181). In applying biological ideas to the urban landscape, they contended that ideas of competition, dominance, invasion and succession would be played out in the city as people competed for space leading to segregation and zonal or sectoral patterns of urban development.

One of the chief legacies of the Chicago School has been the many models of urban land use and social structure that emerged from their 'fastidious empirical research involving intensive qualitative fieldwork and observation' (Hubbard, 2006: 25). These models made generalizations of the internal structure of the city and sought to emphasize significant spatial relationships between the broad land use zones in a town. They also attempted to identify certain spatial characteristics which are common to all towns and provide some understanding of the processes which bring these about (Daniel and Hopkinson, 1989: 117). It was on the basis of the tendency for ecological processes to sort similar households that E.W. Burgess, one of the leading scholars of the Chicago School, formulated one of the first urban models in 1925. His concentric zone model of socio-spatial organization offered a 'neat recapitulation of his theory that as cities grew spatially, patterns of land use would reflect the successive phases of invasion and occupation' (Hubbard, 2006: 27; Burgess, 1925, 1927). Burgess's zonal model was underpinned by the idea that humans adopted a sorting behaviour of competition and relied on an idea of invasion and succession. Based on a detailed analysis of the occurrence of crime and vice in Chicago, he divided the city into five concentric zones: the central business district (CBD); a zone in transition where newly arrived migrants would seek lodging; a zone of working-class housing; a settled residential zone; and finally a ring of commuter suburbs. Burgess argued that there was a positive correlation between the socio-economic status of particular residential areas and their distance from the CBD. According to his model, the outer zones are populated by wealthier residents, who are better able to pay for more desirable spaces, while the less well off are relegated to those less prestigious, mixed zones closer to the CBD.

The concentric zone model effectively provided a description of social segregation in Chicago, based as it was on the analysis of a city with a large population undergoing rapid expansion due to immigration. It was formulated on the basis of a very particular set of economic and political circumstances, however, and subsequent models of urban land use attempted to provide a more refined and widely applicable model (Pacione, 2005: 143). Homer Hoyt, for example, presented his sector model of the city in 1939, based on the detailed analysis of housing rental and value data for over 200 cities (Hoyt, 1939). Hoyt's model demonstrated that urban land values actually tend to vary *within* concentric zones and he presented instead a series of mixed land use sectors developing around the city centre and expanding in particular directions. Both Burgess's and Hoyt's models were later incorporated into Harris and Ullman's multiple-nuclei model of urban land use which attempted to address the excessive simplicity of its predecessors (Harris and Ullmann, 1945; Harris, 1997). Their model drew attention to the multi-nodal character of growth in urban areas and showed instead that many large cities tend not to grow around a single CBD, but are actually organized around a multiplicity of nodal points or nuclei which are themselves controlled by a number of key factors. The multiple-nuclei model attempted to provide a more 'real world' depiction of land use organization in the urban domain and also underscored the significance of specific local contexts in the development of urban spaces.

Conclusion

Although many of these models of urban land use, along with many of those that were developed later in the twentieth century, have to a large degree entered the historiography of urban studies, these early generalizations nonetheless marked an important attempt to make sense of the internal structure of cities and they remain useful pedagogic devices against which to test real-world cities (Pacione, 2005: 144). Furthermore, as originating points for positivist areal studies and factorial ecology, they provided inspiration for a wide range of post-war attempts to map and model patterns of urban land use which 'became an important tradition, and one of the clearest manifestations of geographers' desire to restyle their discipline as a spatial science' (Hubbard, 2006: 28). More recent work in this field has sought to refine traditional models in order to provide concepts of more direct relevance to contemporary urban society (Pacione, 2005: 145). Ultimately, the classical models of urban land use, together with those more recent modifications, provide us with some useful insights into the structure of cities. It is also the case, however, that they could only ever serve as crude simplifications of reality which provide little by way of insight into the actual form or morphology of urban spaces. Their focus on the economic factors at work in creating particular land use regions deflects attention away from the built fabric, historical context and some of the finer processes at work in actually shaping the urban landscape. And so in the next chapter I want to turn to an alternative approach to making sense of the city and one which is more readily located within the realm of historical geography, namely the urban morphology approach that was pioneered by M.R.G. Conzen. In many ways this approach to understanding the city challenges the idea that urban environments develop in a chaotic or vaguely organic manner by focusing instead on a thorough understanding of the structures and processes embedded in urbanization and by paying special attention to how the physical form of a city changes over time and in comparative context.

KEY POINTS

- Over the course of the second half of the twentieth century our world has been transformed into a predominantly urban domain, characterized by an overwhelmingly urban population residing in giant cities.
- Urbanization is the term used to describe an increase in the proportion of people living in urban areas compared to rural areas and in Britain urbanization was spurred on by the industrial revolution.
- The development of cities and the speedy trend towards global urbanization has provided a rich vein of research for human geographers.
- The Chicago School of Urban Ecology played a leading role in formulating models of locational analysis, spurred on by the growth and industrialization of US cities in the 1920s.
- One of the chief legacies of the Chicago School has been the many models of urban land use and social structure that emerged out of their research.

References

Badcock, B. (2002) *Making Sense of Cities*. London: Arnold.

Bulmer, M. (1984) *The Chicago School of Sociology; Institutionalisation, Diversity and the Role of Sociological Research*. Chicago: University of Chicago Press.

Burgess, E.W. (1925) 'The growth of the city: an introduction to a research project', in R.E. Park, E.W. Burgess and R.D. McKenzie (eds) *The City*. Chicago: University of Chicago Press, pp. 47–62.

Burgess, E.W. (1927) 'The determination of gradients in the growth of the city', *Publications of the American Sociological Society*, 21: 178–184.

Daniel, P. and Hopkinson, M. (1989) *The Geography of Settlement*. Harlow: Oliver & Boyd.

Harris, C.D. (1997) '"The nature of cities" and urban geography in the last half-century', *Urban Geography*, 18: 15–35.

Harris, C.D. and Ullmann, E.L. (1945) 'The nature of cities', *Annals of the American Academy of Political and Social Science*, 242: 7–17.

Hoyt, H. (1939) *The Structure and Growth of Residential Neighborhoods in American Cities*. Washington, DC: Federal Housing Administration.

Hubbard, P. (2006) *City*. London: Routledge.

Johnston, R.J. (1971) *Urban Residential Patterns: An Introductory Review*. London: George Bell.

Kingsnorth, P. (2005) *Your Countryside, Your Choice*. London: Campaign to Protect Rural England.

Lawton, R. (ed.) (1989) *The Rise and Fall of Great Cities: Aspects of Urbanisation in the Western World*. London: Belhaven Press.

Lawton, R. and Pooley, C.G. (1992) *Britain. An Historical Geography 1740–1950*. London: Arnold.

Pacione, M. (2005) *Urban Geography: A Global Perspective*. London: Routledge.

FURTHER READING

Badcock, B. (2002) *Making Sense of Cities*. London: Arnold.

Hubbard, P. (2006) *City*. London: Routledge.

Lawton, R. (ed.) (1989) *The Rise and Fall of Great Cities: Aspects of Urbanisation in the Western World*. London: Belhaven Press.

Pacione, M. (2005) *Urban Geography: A Global Perspective*. London: Routledge.

12 GEOGRAPHIES OF URBAN MORPHOLOGY

Yvonne Whelan

Introduction

The study of human settlement, urbanization and the development of cityscapes is, rather like settlement itself, a sprawling topic, that ranges over a vast geographical and conceptual terrain. As a physical reflection of the social organization of space, human settlements have long fascinated geographers, along with scholars in allied fields, like archaeology and history, who have explored the origins of particular forms of settlement through time and across space. Take a city like Dublin, for example, with a settlement history that stretches back to an ancient Gaelic site on the banks of the River Liffey. Geographers have traced the successive waves of human settlement that have shaped this city, focusing attention on the formative role of first the Vikings and then the Anglo-Normans in colonizing this defensible, proto-urban site from the ninth through to the twelfth centuries, as well as on the subsequent phases of both development and decline that characterized the Middle Ages (Simms, 1979). In dissecting the processes at work in shaping this city, they have shown that what was once a relatively small but well-sited proto-urban settlement gradually evolved into a capital city and urban hub in the eighteenth century (Brady and Simms, 2001). When it comes to understanding, describing and classifying landscape change and urban land use, geographers have adopted a number of different explanatory approaches. We could ponder, for example, the various attempts to model the internal organization and land use within cities which emerged from the scholarship of the

Chicago School in the early twentieth century (see Chapter 11). Alternatively, we could look to the scholarship of Carl Sauer with its emphasis on the description and classification of cultural landscapes, or to the ways in which urban spaces can be approached as socially significant and symbolic entities that feed into narratives of identity and shape national iconographies (see Chapter 13). In this chapter I want to consider an approach to the study of the urban landscape that was introduced to English-speaking urban historical geography by M.R.G. Conzen in the 1960s. Conzen's morphological method privileges a more specific concern with the built fabric and the urban morphology or form of the city. His research has provided us with a series of important concepts and methodological techniques that help to explain the geographical character of towns and to chart phases of both development and decline.

Historical Geographies of Urban Morphology

In simple terms, urban morphology refers to the form, arrangement and layout of buildings and the use of land in an urban setting, but it is also concerned with identifying pattern and order with a view to deciphering distinct morphological zones. Urban morphologists are chiefly interested in the evolution of the city from its formative years to its subsequent transformations, and the focus of research in this field rests largely on the tangible results of social and economic forces (Moudon, 1997). Research in urban morphology has a long-standing tradition in Central European geography, with important schools of thought extant in France and Italy (Darin, 1998; Marzot, 2002). The British school of urban morphology is dominated by the work of M.R.G. Conzen (1907–2000) and his technique of town-plan analysis which is based on a close reading of the town plan, building fabric and land use (Whitehand, 1981, 2001). Conzen first came to England as a refugee from Hitler's Germany in 1933 and having qualified as a town planner eventually took up an appointment as a lecturer in the Geography Department at Newcastle University (Slater, 1990). It was there that he began working on what would be his seminal text on the medieval town of Alnwick, Northumberland (Conzen, 1960). This was to be the first in a series of scholarly studies, chiefly on medieval towns in England and Wales, in which he advanced his theories of urban morphology and town-plan analysis (see also Conzen, 1958, 1962, 1988).

Although perhaps somewhat incongruous amidst the disciplinary context of Geography as it was developing during the quantitative revolution, Conzen carved out a school of thought based on rigorous, empirically grounded research which provided methodological and conceptual tools for understanding the historic fabric of urban spaces. Underlying Conzen's approach was his belief that the urban landscape was constantly being changed and adapted over time in order to meet contemporary needs. He argued that the almost forensic examination of the evolution of the urban landscape would shed considerable light on the processes involved. He wrote of the townscape as 'a kind of palimpsest on which the features contributed by any particular period may have been partly or wholly obliterated by those of a later one through the process of site succession or in some other way' (Conzen, 1968: 116). In order, therefore, to explain more fully the origins and development of the urban landscape, Conzen advocated the close analysis of town plans or large-scale maps. Town plans, he argued, enabled the geographer to elaborate on the history of urban spaces especially in the absence of documentary evidence. He devised a methodology called 'town-plan analysis' in order to examine the creation and subsequent transformation of urban form, especially with regard to the spatial development of medieval urban landscapes. In a chapter entitled 'The use of town plans in the study of urban history' published in 1968, Conzen outlined in broad terms the nature and scope of this approach, suggesting at the outset that 'urban history would gain greatly if its practitioners would study town plans as a matter of course' (1968: 115). Town plans constituted one of the three key 'form categories' (to use Conzen's terminology) along with the built fabric (including the building material and architectural style) and the pattern of urban land utilization (1968: 116). He further subdivided the town plan into three constituent parts or elements, namely streets and their arrangement in a street system, plots and their aggregation in street blocks, and the block-plans of buildings (Conzen, 1968: 117).

In his research, Conzen demonstrated that the detailed analysis of the town plan, building fabric and land use, could reveal distinct morphological periods which represented phases of growth and development in urban spaces. He also showed that the three chief elements of the urban landscape reacted at very different rates to agents of change. Thus, he revealed that land use is the most susceptible to change; buildings are adaptable to alternative uses without being physically replaced and so change occurs at a slower rate than land use; and finally the town plan or layout is least susceptible to change. Conzen also showed that a series of similar plots provide the basis for identifying what he termed the 'plan unit' of a town and he argued that 'the recognition of plan units depends, among other things, very much on seeing street spaces, plots and buildings in correlation' (Conzen, 1968: 122). So, periods during which particular combinations of building types and street types are created are called morphological periods. Conzen went on to carry out numerous studies of rather small and well-preserved medieval towns in England and Wales, chiefly Ludlow and Conwy, showing all the while that town-plan analysis could offer the urban researcher an important window into phases of urban development (Lilley, 2000: 6) . He also introduced into the lexicon of urban historical geography a number of key concepts which aid the analysis of urban change. The most notable of these was the fringe belt, a concept that refers to a region of mixed land use at the edge of a built-up area which acts as the physical manifestation of periods of slow movement or actual standstill in the outward expansion of the built-up area (Conzen, 1960; Whitehand, 1967). He also proposed the idea of the burgage cycle which highlights the ways in which land use on a single plot develops over time. As Whitehand argues (2001: 104) 'it was the concepts that he developed about the process of urban development that did most to stimulate a school of thought founded on his work'.

The techniques of town-plan analysis that were pioneered by Conzen have subsequently been applied by geographers in a wide range of historical and geographical settings. As Lilley points out, 'geographers, archaeologists and historians in Britain and Ireland have shown an interest in using settlement morphology as a basis for mapping the historical evolution of medieval urban landscapes' (Lilley, 2000: 5). The Urban Morphology Research Group at the University of Birmingham, which was set up in 1974, continues to be 'an unusually strong centre of research, complementing mainstream traditions in geography' (Moudon, 1997: 4). In recent decades urban morphologists have continued to be actively involved in town-plan analysis, especially on medieval city settlements (Lilley, 2000), but so too have they expanded and advanced the field of urban morphology so that it now encompasses research on a wider range of areas related to cities, such as urban conservation, the management of historic and contemporary townscapes, agents of change in urban and residential locations and suburbanization (Whitehand, 1992; Whitehand and Larkham, 1992). Indeed, in many ways the actual and threatened loss of historical fabric that has accompanied so many urban renewal initiatives has fostered a much deeper appreciation of morphological form and historic urban fabric. One important international and comparative project which focuses particular attention on the form of towns and which makes extensive use of large-scale town plans is the International Historic Towns Atlas Project. In Ireland, for example, this project has resulted in the publication of some 25 studies of Irish towns, each of which charts in meticulous detail the physical growth and development of urban areas, as well as allowing for comparative study (Simms et al., 1986; Simms and Kealy, 2007).

Conclusion

Although Conzen's urban morphological approach to the city has been somewhat marginalized in contemporary human geography, it continues to offer urban historical geographers an important set of methodological and conceptual tools for mapping out the finer geographies of urban land use and tracing the trajectories of historical development in urban spaces. For many urban historical geographers, however, the cultural turn has opened up an increased emphasis on the representational city. As Simms explains, one of the key conceptual developments that enabled and enlivened such research has been the theorization of the urban landscape as a multi-faceted symbolic text to be interrogated and decoded: 'The intention is to understand the iconography of the landscape for what it can tell us about the politically, economically and culturally dominant group in society [...] settlement is both medium and message, site and symbol, terrain and text' (Simms, 2000: 237). So, although the geography of urban settlement continues to fascinate geographers and remains an important focus of research:

> Shifts in social theory and a burgeoning of academic interest in the city have also worked to expand the range of disciplines and perspectives scrutinising the urban setting. The city as an object of analysis has been unbound. Long the concern of a range of interconnected disciplines within

> the social sciences (e.g. geography and urban planning), the city is now open to the distinctive approaches of those working within interdisciplinary fields such as cultural and feminist studies. (Jacobs, 1993: 827)

A body of work has subsequently evolved which approaches urban spaces as sites of representation and as dynamic symbolic constituents of socio-cultural and political systems. Much of this research is sensitive to the symbolism and ideological significance of the built form in the construction, mobilization and representation of identity, and foregrounds the links between ideology, power and the built form (see Section 5).

KEY POINTS

- Geographers have adopted a range of different approaches in order to make sense of urban land use.
- Urban morphology refers to the form, arrangement and layout of buildings and the use of land and is also concerned with identifying pattern and order with a view to identifying distinct morphological zones.
- The British school of urban morphology was pioneered by M.R.G. Conzen and his technique of town-plan analysis is based on a close reading of the town plan, building fabric and land use.
- For many urban historical geographers the cultural turn has opened up an increased emphasis on the representational city and foregrounds the links between ideology, power and the built form.

References

Brady, J. and Simms, A. (eds) (2001) *Dublin through Space and Time*. Dublin: Four Courts.

Conzen, M.R.G. (1958) 'The growth and character of Whitby', in G.H.J. Daysh (ed.) *A Survey of Whitby and the Surrounding Area*. Eton: Shakespeare Head Press, pp. 39–49.

Conzen, M.R.G. (1960) *Alnwick, Northumberland: A Study in Town-Plan Analysis*, Institute of British Geographers Publication, 27. London: George Philip.

Conzen, M.R.G. (1962) 'The plan analysis of an English city centre', in K. Norborg (ed.) *Proceedings of the IGU Symposium in Urban Geography Lund 1960*. Lund: Gleerup.

Conzen, M.R.G. (1988) 'Morphogenesis, morphological regions and secular human agency in the historic townscape, as exemplified by Ludlow', in D. Denecke and G. Shaw (eds) *Urban Historical Geography: Recent Progress in Britain and Germany*. Cambridge: Cambridge University Press, pp. 252–272.

Darin, M. (1998) 'The study of urban form in France', *Urban Morphology*, 2: 63–76.
Jacobs, J.M. (1993) 'The city unbound: qualitative approaches to the city', *Urban Studies*, 30: 827–848.
Lilley, K.D. (2000) 'Mapping the medieval city: plan analysis and urban history', *Urban History*, 27(1): 5–30.
Marzot, N. (2002) 'The study of urban form in Italy', *Urban Morphology*, 6: 59–73.
Moudon, A.V. (1997) 'Urban morphology as an emerging interdisciplinary field', *Urban Morphology*, 1: 3–10.
Simms, A. (1979) 'Medieval Dublin: a topographical analysis', *Irish Geography*, 12: 25–41.
Simms, A., Clarke, H.B. and Gillespie, R. (eds) (1986) *Irish Historic Towns Atlas*. Dublin: Royal Irish Academy.
Simms, A. (2000) 'Perspectives on Irish settlement studies', in T. Barry (ed.) *A History of Settlement in Ireland*. London: Routledge, pp. 228–247.
Simms, A. and Kealy, L. (2007) 'The study of urban form in Ireland', *Urban Morphology*, 12(1): pp. 37–45.
Slater, T.R. (1990) 'Starting again: recollections of an urban morphologist', in T.R. Slater (ed.) *The Built Form of Western Cities: Essays for M.R.G. Conzen on the Occasion of his Eightieth Birthday*. Leicester: Leicester University Press, pp. 3–22.
Whitehand, J.W.R. (1967) 'Fringe belts: a neglected aspect of urban geography', *Transactions of the Institute of British Geographers*, 41: 223–233.
Whitehand, J.W.R. (1981) 'Background to the urban morphogenetic tradition', in J.W.R. Whitehand (ed.) *The Urban Landscape: Historical Development and Management*. London: Academic Press, pp. 1–24.
Whitehand, J.W.R. (1992) 'Recent advances in urban morphology', *Urban Studies*, 29: 619–636.
Whitehand, J.W.R. (2001) 'British urban morphology: the Conzenian tradition', *Urban Morphology*, 5: 103–109.
Whitehand, J.W.R. and Larkham, P.J. (1992) (eds) *Urban Landscapes: International Perspectives*. London: Routledge.

FURTHER READING

Conzen, M.R.G. (1968) 'The use of town plans in the study of urban history', in H.J. Dyos (ed.) *The Study of Urban History*. London: Arnold, pp. 13–30.
Lilley, K.D. (2000) 'Mapping the medieval city: plan analysis and urban history', *Urban History*, 27(1): 5–30.
Slater, T.R. (ed.) (1990) *The Built Form of Western Cities: Essays for M.R.G. Conzen on the Occasion of his Eightieth Birthday*. Leicester: Leicester University Press, pp. 3–22.
Whitehand, J.W.R. (2001) 'British urban morphology: the Conzenian tradition', *Urban Morphology*, 5: 103–109.

Section 5 Place and Meaning

13 LANDSCAPE AND ICONOGRAPHY

Yvonne Whelan

Introduction: Understanding Landscape

> A landscape is a cultural image, a pictorial way of representing, structuring or symbolising surroundings. [...] A landscape park is more palpable but no more real, no less imaginary, than a landscape painting or poem. [...] And of course, every study of a landscape further transforms its meaning, depositing yet another layer of cultural representation. (Cosgrove and Daniels, 1988: 1)

Landscape holds a central place in geographical inquiry and approaches to its study and analysis have evolved over time. This chapter begins by reflecting on the development of the landscape concept in human geography, before considering more closely the approaches to landscape study that developed out of the 'new cultural geography' in the 1990s. In particular, it examines the symbolic geography of the cultural landscape and the rich nexus of inquiry that has coalesced around landscape, memory and identity. Historical geographers have long been concerned with the study of landscape in all of its various forms, from the cultural to the physical, the rural to the urban, and they have followed a variety of paradigms in order to represent its complexity. In the late nineteenth century, for example, the term *landschaft* came into common parlance in German geography to describe the appearance of the earth's surface or a particular region. At the same time in France, geographers approached landscape or *paysage* as 'an expression of human activity, as a human imprint upon the land' (Baker, 2003: 110). In many ways this work foreshadowed the influential research of Carl Sauer that would follow in the 1920s. When Carl Sauer, a Professor of Geography at the University of California at Berkeley, introduced the concept into American cultural geography, he drew to some extent on the German *landschaft* tradition and approached landscape as the 'unit concept' of geography. For Sauer and his many students at Berkeley, landscape was an object to be studied. He was centrally concerned with how natural landscapes were transformed into cultural landscapes and in his 1925 paper 'The morphology of landscape' argued that:

> the cultural landscape is fashioned from a natural landscape by a culture group. Culture is the agent, the natural area is the medium, and the cultural landscape is the result. Under the influence of a given culture, itself changing though time, the landscape undergoes development, passing through phases, and probably reaching ultimately the end of its cycle of development. With the introduction of a different, that is alien culture, a rejuvenation of the cultural landscape sets in, or a new landscape is superimposed on the remnants of an older one. The natural landscape is of course of fundamental importance, for it supplies the material out of which the cultural landscape is formed. The shaping force, however, lies in the culture itself. (Sauer, 1963 [1925]: 343)

At the heart of Sauer's scientific endeavour, therefore, was a concern with the morphology of landscape, and more particularly with the ways in which people (or 'culture groups') left their mark on the landscape. Sauer's research interests ranged widely, but a key strand focused on the human forces that shaped landscapes and he placed particular emphasis on the material artefacts, such as house types, that defined specific regions, especially in pre-European America. Although his legacy in cultural geography is extensive, his conceptualization of landscape and culture has also been controversial (Cosgrove and Jackson, 1987; Gregory and Ley, 1988; Price and Lewis, 1993). His accounts of seemingly obvious, tangible, countable and mappable phenomena evident in the landscape were later criticized for resulting in 'endless studies of house-types, field patterns, log-cabin construction methods, and place-imagery in music lyrics', and for creating a sub-field of cultural geography that was 'antiquarian, particularistic and socially irrelevant' (Mitchell, 2000: xiv; see also Cosgrove and Jackson, 1987; Kong, 1997).

In the middle of the twentieth century two scholars, W.G. Hoskins and J.B. Jackson, exerted some enduring influences on the study of landscape. Hoskins, an English landscape historian, authored his pioneering text on landscape history, *The Making of the English Landscape* in 1955 (Matless, 1993). Jackson, an American geographer, founded *Landscape* magazine in 1951 and placed particular emphasis on the study of vernacular landscapes and the cultural symbolism of landscape. Both Hoskins' and Jackson's work resonated in many ways with that of Sauer, especially given their close attention to landscape as a material entity which should be studied closely at first-hand. But they also exerted new influences and together with Sauer set the intellectual agenda for landscape studies up to the late 1970s. The publication of Donald Meinig's edited collection *The Interpretation of Ordinary Landscapes* in 1979 marked something of a watershed moment for landscape studies in human geography. The essays in this book demonstrated in many ways Sauer's continuing legacy but they also pointed to new directions which would take hold, directions that would trumpet notions of landscape as 'text' and 'symbol'. As geographers became increasingly influenced by theoretical developments in the humanities and social sciences more broadly, conceptualizations of landscape and culture were reformulated in the 1980s. Out of this intellectual ferment emerged a diverse range of new cultural geographies which critiqued more traditional understandings of culture and landscape.

New Cultural Geographies of Landscape

As space became a resurgent analytic in the late 1980s and 1990s, geographers began to revise received views of the cultural landscape. Instead of seeing landscapes as sites of settlement, they sought 'to recover layers of meaning lying beyond (or "beneath") those surface remains and relict features' (Gregory, 1994: 145). The work of geographers like Denis Cosgrove, Stephen Daniels and James Duncan was especially influential in this regard. In the seminal *Social Formation and Symbolic Landscape*, published in 1984, Cosgrove conceptualized landscape as 'an ideologically-charged and very complex cultural product' and demonstrated a reflexive, sympathetic approach to the nuances and tensions evident in the processes of landscape creation (1984: 11). His fresh theorization of landscape also underscored the very situatedness of the geographical researcher. In their edited collection *The Iconography of Landscape* (1988) Cosgrove and Daniels advanced the idea of landscape as 'a cultural image, a pictorial way of representing, structuring or symbolising surroundings' (1988: 1). Making use of an iconographic approach borrowed from art history, Cosgrove and Daniels focused attention on the symbolic make-up of cultural landscapes and on how landscapes might embody relations of power that are socially negotiated and tested. The iconographic method had originally been employed by art historians to 'probe meaning in a work of art by setting it in its historical context and, in particular, to analyse the ideas implicated in its imagery. [It] consciously sought to conceptualise pictures as encoded texts to be deciphered by those cognisant of the culture as a whole in which they were produced' (Cosgrove and Daniels, 1988: 2). The adaptation of this approach for the study of the cultural landscape demanded that geographers focus renewed attention on the symbolic geographies embedded in the construction and design of landscapes and look beneath the surface to reveal the politics, processes and symbolic qualities written into it (Robertson and Richards, 2003: 4). James Duncan, meanwhile, was instrumental in positing a view of landscape as text and as 'one of the central elements in a cultural system, [...] a signifying system through which a social system is communicated, reproduced, experienced and explored' (Duncan, 1990: 184).

Metaphors of landscape as text, symbol and 'way of seeing' were richly suggestive of a more overtly interpretative set of approaches to the study of landscape. Many of the 'new cultural geographers' of the 1990s were especially attentive to the idea of landscape as a construction and site of emblematic representation, a 'particular way of composing, structuring and giving meaning to an external world whose history has to be understood in relation to the material appropriation of land' (Cosgrove and Jackson, 1987: 96). The idea of landscape as an ideological and dynamic entity, one that could 'endorse, legitimise, and/or challenge social and political control' (Kong, 1993: 24) paved the way for a whole range of innovative readings of symbolic city spaces, many of which emphasized the politics of landscape and the significance of public monuments in particular in contributing to the symbolic landscape and cultivating identity. The cultural landscape came to be conceptualized as a 'whole mix of human activities, systems of meaning and symbolic forms within the physical setting of the city' (Logan, 2000: 1). Seemingly ordinary, everyday spaces were seen to 'reveal, represent and symbolise the relationships of power and control out of which they have emerged and the human processes that have transformed and continue to transform them' (Robertson and Richards, 2003: 4). One especially fruitful area of landscape inquiry coalesced around the nexus of landscape, memory and identity.

Landscape, Memory and Identity

An important research theme which emerged in the mid-1990s placed particular emphasis on iconic landscapes of national identity (Wylie, 2007: 192). Many cultural and historical geographers set about probing the politics of public memory in a wide range of geographical settings and historical contexts, elaborating on the symbolic geographies of what is remembered in the public domain, but also upon that which is forgotten (Foote and Azaryahu, 2007). As focal points of public memory, monuments have been shown to possess an enduring significance as identity resources and as spaces 'where memories converge, condense, conflict, and [which] define relationships between past, present, and future' (Davis and Starn, 1989: 3). The trajectories of memory work in historical and cultural geography have been many and varied, but an especially rich seam has exposed the ways in which monuments act as mnemonic devices that feed into constructions of identity and articulate strategies of resistance (Johnson, 1994, 1995; Heffernan, 1995; Auster, 1997; Atkinson and Cosgrove, 1998; Jezernik, 1998; Osborne, 1998; Cooke, 2000; Hoelscher and Alderman, 2004). A large body of memory work has emerged which foregrounds the ways in which 'memorials and monuments are political constructions, recalling and representing histories selectively, drawing popular attention to specific events and people and obliterating or obscuring others' (Frances and Scates, 1989: 72).

Some of these ideas can be fleshed out by looking at the historical geography of Dublin city. Ireland has long provided a rich resource base for work on the overlapping relationships between landscape, memory and identity, especially given the contested political relationship between Britain and Ireland which placed significant demands upon the symbolic fabric of the cultural landscape. By trawling through archives and uncovering information relating to the development of the city before and after the achievement of political independence from Britain in 1922 one can identify the ways in which the symbolic fabric of the city changed dramatically. Statues, street names, public buildings and urban planning projects each served as sites of memory and meaning, demonstrating the powerful role played by the cultural landscape in articulating the geographies of political and cultural identity. So, those seemingly innocuous aspects of the landscapes in which we live, whether it be a monument at the end of the road or the name attached to a street, are anything but accidental or ornamental features. On the contrary, they are very often the signifying threads that make up the fabric of our cultural landscapes and in turn help to construct and consolidate narratives of identity at a variety of spatial scales.

If we zone in on one street in particular, O'Connell Street, Dublin's central thoroughfare, we can identify some of the ways in which monuments make meaning in public space. We can also see just how fluid and subject to change those meanings are, in a manner which reinforces Dwyer's point that monuments need

> to be conceived of as in the process of becoming instead of existing in a static, essentialized state. Rather than possessing a fixed, established meaning, monuments are momentarily realized in a nexus of social relations as the result of attempts to define the meaning of representations, which nevertheless remain open to dispute and change. (2004: 425; see also Dwyer, 2002)

Before Ireland achieved independence from Britain, the centre of this street was dominated by a monument dedicated to Lord Nelson which had been unveiled in 1809, on the anniversary of the Battle of Trafalgar. This monument, alongside many others throughout the city dedicated to members of the British military and monarchy, stood as a symbol of Ireland's status as a city of the British Empire (Whelan, 2001a, 2001b). But Dublin was also a contested space and, even before the achievement of independence, monuments dedicated to a swathe of figures broadly related to Irish nationalist politics were used to 'challenge in stone' British rule in Ireland. This process gathered momentum when Ireland became independent and memorial spaces became an important tool in the nation-building agenda. It is also notable that many older monuments of empire were destroyed during this period, underlining the significance of forgetting and erasure in the process of building nations (Whelan, 2002). So, the jubilant scenes that had prevailed in 1809 when Nelson's Pillar was unveiled stand in marked contrast to events that took place some years later in 1966. In the early morning of 8 March a bomb that had been planted by a republican splinter group at the base of Nelson's Pillar went off, badly damaging the column and the statue of Nelson. The site remained vacant until December 2000 when the Irish government announced that a new monument known

Figure 13.1 The Spire of Dublin, O'Connell Street (Photo: Yvonne Whelan)

as the 'Dublin Spire' was to be erected on the site of the old pillar (Figure 13.1). This slender steel structure, 120 metres high and three metres in diameter at its base, was constructed 'in celebration of Ireland's confident future in the third millennium' and contrasts markedly with those earlier erected monuments of both empire and nation-building (Ritchie, 2004).

In many ways the story of Nelson's Pillar is illustrative of the broader narrative of Dublin's iconography as it evolved before and after the creation of the Irish Free State in 1922. It also reinforces the point that 'literally and figuratively monuments and memorials set dominant socio-spatial relations "in stone"' (Hubbard et al., 2003: 150). Like other landscape elements, statues and public monuments demonstrate and reaffirm the power of dominant groups or individuals. They are, after all, 'designed and planned, with all of the narrative choices and biases this entails, by those who have the time, resources, and, most importantly, the state mandate to define the past' (Dwyer, 2002: 32). But so too can monuments stand as focal points of resistance. As the brief allusion to Nelson's Pillar illustrates, the meaning attached to memorial sites is not fixed, but rather fluid and malleable and subject to change as social, cultural and political contexts evolve or rupture. As Baldwin et al. point out: 'Monuments are also the sites of resistance to the meanings that they try to fix, and this resistance may come from various quarters, conservative and radical. Indeed, it is the power of monuments as symbols that sets them up as sites where challenges to authority can strike at what the powerful hold most dear' (1999: 255).

Conclusion

The new directions in landscape study that were charted in the 1990s spawned a rich strand of research relating to landscape and its ideological and symbolic meanings. In the years that have followed, the landscape concept has continued to evolve, and ideas of landscape as text, symbol and 'way of seeing' have been subject to sustained critique on a range of different grounds, and especially from feminist and Marxist scholars (Rose, 1993; Mitchell, 2000; see also Wylie, 2007). Today, landscape remains at the heart of a great deal of geographical research in both human and physical geography and it continues to be conceptualized in new and innovative ways, while also maintaining a large degree of continuity with those earlier, more morphological approaches. Recent landscape studies have broadened out substantially to include more probing analyses of gender, sexuality, race, diaspora and the body. Mitchell's idea of landscape as a 'vortex' is perhaps a useful note to end on, for it draws attention to the turbulent nature of the landscape terrain, 'within which swirl all manner of contests – between classes, over gender structures, around issues of race and ethnicity, over meaning and representation, and over built form and social use. The landscape serves all at once as mediator, integrator, and actor in these struggles' (2000: 139). For many historical and cultural geographers, research on landscape is now closely informed by a range of fresh theoretical insights, from postcolonialism to phenomenology, and from post-structuralism to non-representational theory.

KEY POINTS

- Historical geographers have long been concerned with the study of landscape in all of its various forms, from the cultural to the physical, the rural to the urban, and have followed a variety of paradigms in order to represent its complexity.
- Carl Sauer, a Professor of Geography at the University of California at Berkeley, was instrumental in introducing the landscape concept into American cultural geography in the twentieth century. Sauer was centrally concerned with the human forces that shaped landscapes and the material artefacts that defined specific regions.
- As geographers became increasingly influenced by theoretical developments in the humanities and social sciences in the 1980s, the conceptualization of landscape changed. Interpretative metaphors of landscape as text, symbol and 'way of seeing' came to prominence as geographers set about conceptualising landscape as an ideological construction and site of emblematic representation.
- An important research theme which emerged in the mid-1990s placed particular emphasis on iconic landscapes of national identity. One especially fruitful area of landscape inquiry coalesced around the nexus of landscape, memory and identity.
- Today, landscape remains at the heart of a great deal of geographical research in both human and physical geography and it continues to be conceptualized in new and innovative ways, closely informed by a range of fresh theoretical insights.

References

Atkinson, D. and Cosgrove, D. (1998) 'Urban rhetoric and embodied identities: city, nation, and empire at the Vittorio Emanuele II Monument in Rome', *Annals of the Association of American Geographers*, 88(1): 8–49.

Auster, M. (1997), 'Monument in a landscape: the question of "meaning", *Australian Geographer*, 28(2): 219–227.

Baker, A.R.H. (2003) *Geography and History: Bridging the Divide*. Cambridge: Cambridge University Press.

Baldwin, E., Longhurst, B., McCracken, S., Ogborn, M. and Smith, G. (1999) *Introducing Cultural Studies*. London: Prentice Hall.

Cooke, S. (2000) 'Negotiating memory and identity: the Hyde Park holocaust memorial, London', *Journal of Historical Geography*, 26(3): 449–465.

Cosgrove, D. (1984) *Social Formation and Symbolic Landscape*. London: Croom Helm.

Cosgrove, D. and Daniels, S. (eds) (1988) *The Iconography of Landscape*. Cambridge: Cambridge University Press.
Cosgrove, D. and Jackson, P. (1987) 'New directions in cultural geography', *Area*, 19(2): 95–101.
Davis, N.Z. and Starn, R. (1989) 'Introduction: memory and counter-memory', *Representations*, 26(1): 1–6.
Duncan, J.S. (1990) *The City as Text: the Politics of Landscape Interpretation in the Kandyan Kingdom*. Cambridge: Cambridge University Press.
Dwyer, O.J. (2002) 'Location, politics, and the production of civil rights memorial landscapes', *Urban Geography*, 23(1): 31–56.
Dwyer, O.J. (2004) 'Symbolic accretion and commemoration', *Social and Cultural Geography*, 5(3): 419–435.
Foote, K.E and Azaryahu, M. (2007) 'Towards a geography of memory: geographical dimensions of public memory and commemoration', *Journal of Political and Military Sociology*, 35(1): 125–144.
Frances, R. and Scates, B. (1989) 'Honouring the Aboriginal dead', *Arena*, 86: 72–80.
Gregory, D. (1994) *Geographical Imaginations*. Oxford: Blackwell.
Gregory, D. and Ley, D. (1988) 'Culture's geographies', *Environment and Planning D: Society and Space*, 6(1): 115–116.
Heffernan, M. (1995), 'For ever England: the western front and the politics of remembrance in Britain', *Ecumene*, 2(3): 293–323.
Hoelscher, S.D. and Alderman, D.H. (2004) 'Memory and place: geographies of a critical relationship', *Social and Cultural Geography*, 5(3): 347–55.

Hubbard, P., Faire, L. and Lilley, K. (2003) 'Memorials to modernity? Public art in the "city of the future"', *Landscape Research*, 28(2): 147–169.
Jezernik, B. (1998) 'Monuments in the winds of change', *International Journal of Urban and Regional Research*, 22(4): 582–588.
Johnson, N.C. (1994) 'Sculpting heroic histories: celebrating the centenary of the 1789 rebellion in Ireland', *Transactions of the Institute of British Geographers*, 19(1): 78–93.
Johnson, N.C. (1995) 'Cast in stone: monuments, geography, and nationalism', *Environment and Planning D: Society and Space*, 13(1): 51–65.
Kong, L. (1993) 'Political symbolism of religious building in Singapore', *Environment and Planning D: Society and Space*, 11(1): 23–45.
Kong, L. (1997) 'A "new" cultural geography? Debates about invention and reinvention', *Scottish Geographical Magazine*, 113(3): 177–185.
Logan, W.S. (2000) *Hanoi: Biography of a City*. Sydney: University of New South Wales Press.
Matless, D. (1993) 'One man's England: W.G. Hoskins and the English culture of landscape', *Rural History*, 4(2): 187–207.
Meinig, D.W. (ed.) (1979) *The Interpretation of Ordinary Landscapes*. New York: Oxford University Press.
Mitchell, D. (2000) *Cultural Geography: A Critical Introduction*. Oxford: Blackwell.
Mitchell, D. (2003) 'Cultural landscapes: just landscapes or landscapes of justice?', *Progress in Human Geography*, 27(6): 787–796.
Osborne, B.S. (1998) 'Constructing landscapes of power: the Georges Etienne Cartier memorial, Montreal', *Journal of Historical Geography*, 24(4): 431–458.

Price, M. and Lewis, M. (1993) 'The reinvention of cultural geography', *Annals of the Association of American Geographers*, 83(1): 1–17.
Ritchie, I. (2004) *The Spire*. London: Categorical Books.
Robertson, I. and Richards, P. (eds) (2003) *Studying Cultural Landscapes*. London: Hodder Arnold.
Rose, G. (1993) *Feminism and Geography*. Cambridge: Polity Press.
Sauer, C.O. (1963 [1925]) 'The morphology of landscape', in J. Leighly (ed.) *Land and Life: A Selection from the Writings of Carl Ortwin Sauer*. Berkeley: University of California Press, pp. 315–350.
Whelan, Y. (2001a) 'Monuments, power and contested space: the iconography of Sackville Street, Dublin before Independence (1922)', *Irish Geography*, 34(1): 11–33.
Whelan, Y. (2001b) 'Symbolising the state: the iconography of O'Connell Street, Dublin after Independence (1922)', *Irish Geography*, 34(2): 135–156.
Whelan, Y. (2002) 'The construction and destruction of a colonial landscape: commemorating British monarchs in Dublin before and after independence', *Journal of Historical Geography*, 28(4): 508–533.
Wylie, J. (2007) *Landscape*. London: Routledge.

FURTHER READING

Cosgrove, D. (1984) *Social Formation and Symbolic Landscape*. London: Croom Helm.
Cosgrove, D. and Daniels, S. (eds) (1988) *The Iconography of Landscape*. Cambridge: Cambridge University Press.
Duncan, J.S. (1990) *The City as Text: The Politics of Landscape Interpretation in the Kandyan Kingdom*. Cambridge: Cambridge University Press.
Wylie, J. (2007) *Landscape*. London: Routledge.

14 CONCEPTUALIZING HERITAGE

Yvonne Whelan

Introduction

In the autumn of 2007 hundreds of protesters gathered at the Hill of Tara in County Meath, a prehistoric site of Irish cultural heritage and one of Ireland's most venerated archaeological sites. There they assembled in the form of a giant human harp, not as part of an elaborate stunt to gain entry into the *Guinness Book of Records*, but instead to promote the burgeoning campaign to re-route the national M3 motorway. Designed to ease the traffic in towns along the Meath corridor which have become part of the Dublin commuter belt, the M3 would cut through the Tara-Skryne Valley, coming within a couple of kilometres of the Tara site. The campaign against the motorway gathered momentum throughout the year as hundreds joined the campaign to stop the destruction of this historic site. The Hill of Tara was subsequently listed among the world's 100 most endangered heritage sites and was added to the crisis list of the World Monument Fund, one of the leading non-governmental organizations for the protection of cultural heritage. This clash between the commercially minded with their development agenda, and those with a desire to preserve and protect aspects of the ancient past, escalated into a bitter dispute. The episode captures in microcosm a much broader set of debates about cultural heritage and the wider uses and abuses of the past in contemporary contexts. In recent decades, heritage has become a hot topic, one that has sparked the interest of historical and cultural geographers alike. The political, economic and cultural uses of the past in our landscapes of the present have all come under examination in a wide range of geographical contexts and at a variety of scales. But what do we really mean when we talk about heritage? And how have geographers set about conceptualizing it?

Understanding Heritage

> Heritage brings manifold benefits: it links us with ancestors and offspring, bonds neighbors and patriots, certifies identity, roots us in time-honoured ways. But heritage is also oppressive, defeatist, decadent. Miring us in the obsolete, the cult of heritage allegedly immures life within museums and monuments. [...] heritage is also accused of undermining historical truth with twisted myth. (Lowenthal, 1998: xiii–xiv)

The word 'heritage' evokes different things for different people and it is, therefore, notoriously difficult to define. We can quite literally locate it in the landscapes around us. Think, for example, of the 43 celebrated heritage coasts in England and Wales, part of the heritage coast classification scheme that was initiated in 1972 to protect coastline of special scenic and environmental value from undesirable development. Today, some 31 per cent of the coast in England and 42 per cent in Wales is protected under this scheme and many of these coasts are in turn part of larger National Parks or Areas of Outstanding Natural Beauty. These natural, physical landscapes form important components of heritage, and they symbolize our collective inheritance of the natural environment (Aitchison et al., 2000). But heritage also resides in other, more cultural forms such as historic buildings or ancient monuments. The United Nations Educational, Scientific and Cultural Organization (UNESCO) lists more than 800 sites of cultural and natural heritage from 130 countries on its World Heritage List and includes 'places as unique and diverse as the wilds of East Africa's Serengeti, the Pyramids of Egypt, the Great Barrier Reef in Australia and the Baroque cathedrals of Latin America'.[1]

Such tangible forms of cultural heritage contrast sharply with other more *intangible* forms, which have captured the attention of UNESCO in recent years. According to the 2003 Convention for the Safeguarding of the Intangible Cultural Heritage, 'the intangible cultural heritage – or living heritage – is the mainspring of our cultural diversity and its maintenance a guarantee for continuing creativity'.[2] The Convention defines intangible cultural heritage as 'the practices, representations, expressions, as well as the knowledge and skills, that communities, groups and, in some cases, individuals recognise as part of their cultural heritage'.[3] It can be made manifest in a range of ways, such as oral traditions and expressions, including language; performing arts (such as traditional music, dance and theatre); social practices, rituals and festive events; knowledge and practices concerning natures and the universe; and traditional craftsmanship. So, the UNESCO list of intangible cultural heritages ranges from Albanian Folk Iso-Polyphony to the Carnival of Binche in Belgium and the cultural space of the Jemaa el-Fnaa Square in Marrakech, Morocco.[4]

Heritage, then, can be found in a whole range of tangible forms that are rooted in the cultural landscapes in which we live (buildings, statues, monuments, houses, as well as physical landscapes). But heritage can also be intangible with different forms of music, dance, folklore and oral culture comprising significant dimensions of our cultural heritage. What each of these myriad forms of heritage share in common, however, is an underlying emphasis on selective aspects of 'the past' being repositioned or located in the present. As Brian Graham et al. observe:

> heritage is that part of the past which we select in the present for contemporary purposes, be they economic, cultural, political or social. The worth attributed to these artefacts rests less in their intrinsic merit than in a complex array of contemporary values, demands and even moralities. (2000: 17)

Heritage, therefore, can be distinguished from history by virtue of its very present-centredness. It encapsulates a wide variety of landscapes, buildings and aspects of material culture, together with the many centres of heritage interpretation that have evolved in recent years as explanatory sites and as significant sources of income for local communities. At its core heritage is highly selective, culturally constructed and whether tangible or intangible, official or unofficial, fulfils a variety of oftentimes competing functions. What is designated as 'heritage' and worthy of preservation invariably fulfils a particular cultural role. And so heritage sites can serve as icons of identity which feed into constructions of identity and particular narratives of nationhood (MacCannell, 1992). It is also the case, however, as Tunbridge and Ashworth argue, that 'all heritage is someone's heritage and therefore logically not someone else's. [...] any creation of heritage from the past disinherits someone completely or partially, actively or potentially' (1996: 21). Thus, heritage has the power to reinforce hegemonic power and strengthen group identity, while at the same time potentially excluding those who may be less dominant (Aplin, 2002: 16–17).

But heritage is also about economic commodification, a point hinted at by Schouten when he suggests that 'heritage is history processed through mythology, ideology, nationalism, local pride, romantic ideas or just plain marketing, into a commodity' (Schouten, 1995: 21). Those aspects of our cultural heritage that are deemed worthy of preservation may well be endowed with cultural significance, but this often conflicts with the fact that they may also be commodities with significant economic value, especially in the context of international tourism and development (Urry, 1990, 1995; Sack, 1992; Graham and Howard, 2008). Sites of heritage invariably attract large numbers of tourists and generate income, hence the extent to which aspects of our cultural heritage are packaged and marketed for consumers. The heritage industry is a thriving one and heritage tourism is worth a great deal in national and local economies. As sites of cultural and natural heritage are rapidly consumed, however, so too can they be threatened or destroyed. We can get a sense of this potential for conflict by considering the ancient megalithic site of Stonehenge in Wiltshire. Managed by English Heritage, the site receives over 800,000 visitors a year, while a further 200,000 peer through the roadside fence to view it. There is an inevitable tension between the revenue that such numbers generate and the potential threat to the site itself under the sheer weight of visitors. It follows that 'if heritage is to be commodified and commercialised, this must be done with as much sensitivity and care as possible' (Aplin, 2002: 57). Heritage, therefore, plays something of a dual role as locus of both cultural and economic capital and there exists an inevitable tension or 'dissonance' between both domains (Tunbridge and Ashworth, 1996; Landzelius, 2003).

Representing Famine Heritage

Heritage clearly fulfils a range of often competing functions which straddle the economic, cultural and political domains (Graham, 2002). One of these that I would like to explore further here relates to the pivotal role of heritage in shaping narratives of identity. In this context, heritage overlaps quite forcefully with concepts like landscape and memory in refining quite complex collective memories into tangible sites (see Chapter 13). As Lowenthal argues, 'heritage distils the past into icons of identity, bonding us with precursors and progenitors, with our own earlier selves, and with promised successors' (1994: 43). For emigrant communities in particular, the desire to re-imagine spaces of the homeland has engendered a variety of commemorative and heritage-based practices. Migration and displacement sharpens nostalgia and fosters a hunger for heritage, while 'displaced persons are displaced not just in space but in time; they have been cut off from their own pasts. [...] If you cannot revisit your own origins – reach out and touch them from time to time – you are forever in some crucial sense untethered' (Lively, cited in Lowenthal, 1998: 9).

For Irish-Americans, especially those who trace their ancestry to the three-quarters of a million Irish people who made their way to the United States during and after the Great Hunger of the mid-nineteenth century, the commemoration and representation of that past has a great deal of cultural and political significance (Kelleher, 2002). This was brought to the fore in recent years with the sesquicentenary of the Famine in the mid-1990s, an anniversary that marked the onset of a concerted effort by many Irish-Americans to commemorate the plight of their ancestors, through the development of sites of famine heritage and memory. The cultural landscape came to act as an important stage upon which scenes of the Great Irish Famine were represented, and memorials were added to a number of cityscapes, notably Boston, Buffalo, Philadelphia and New York. These sites invariably represented the famine and migration experience as a triumphant journey from poverty to socio-economic success (Graham, 2007; Whelan and Harte, 2007). In their form and composition they projected a rather sanitized version of the past, one which erased the rawness of the Famine trauma (Edkins, 2003: 122). Such sanitized spaces of cultural heritage are grist to the mill of those who suggest that heritage invariably presents a kind of 'bogus history' or twisted myth. As Johnson notes, 'For some, heritage tourism is seen as a form of bogus history that commodifies the past, distorts the real histories conveyed through more conventional means (e.g., academic writing), and biases views of the past by selectively presenting evidence' (1996: 552; see also Hewison, 1987). One site, however, which manages to meld landscape, memory and heritage in a novel and much more challenging way is the Irish Hunger Memorial in Battery Park, City New York.

Designed by New York-based sculptor Brian Tolle, the Irish Hunger Memorial was unveiled in July 2002 in Lower Manhattan by the then President of Ireland, Mary McAleese, in what she described as 'the memory-shadow of that tragic absence that is the Twin Towers of the World Trade Center'.[5] The origins of the project stretch back to March

2000, when the then New York governor, George E. Pataki, launched a competition for a memorial that would 'serve as a reminder to millions of New Yorkers and Americans who proudly trace their heritage to Ireland of those who were forced to emigrate during one of the most heart-breaking tragedies in the history of the world'.[6] When Tolle's design was chosen as winner from the five invited schemes, a site was prepared in Battery Park City overlooking the Hudson River, in view of the Statue of Liberty and Ellis Island. Here, the centrepiece of the memorial was installed: a derelict stone cottage set in a quarter-acre of farmland, transplanted from County Mayo in Ireland, and reconstructed atop a raised and tilted concrete base, which comprises the second element of the memorial (Figure 14.1). The plot was planted with Mayo vegetation and strewn with 32 fieldstones, each one representing an Irish county, and a single carved pilgrim stone. One entry to the base of the memorial takes the viewer through a passageway that is redolent of a neolithic burial mound, the walls of which are lined with glass-covered strands of text that deliberately mingle Famine facts, statistics about world hunger and obesity today and quotations from literature and song (Figure 14.2).

Figure 14.1 Stone cottage at the Irish Hunger Memorial, New York (Photo: Yvonne Whelan)

Figure 14.2 The entrance to the Irish Hunger Memorial, New York (Photo: Yvonne Whelan)

The Irish Hunger Memorial is a poignant and affecting site of cultural heritage. Rather than taking the form of a figurative memorial, one which would be easily readable, the memorial poses a challenge, inviting the visitor into a dialogue with its many composite elements and demanding a level of engagement not often associated with more figurative memorials. In place of sculpted art we have the actual ruins of an Irish cottage, and with that the past is evoked through the very realness of a patch of transplanted Irish earth. Similarly, the quarter-acre plot which makes up the site, replicates the maximum area that farmers could hold in order to qualify for British government aid in the 1840s, while the cottage's roofless state registers the destructive impact of Famine evictions.

Conclusions

With its emphasis on disruption and dislocation the the Irish Hunger Memorial in New York presents a more complex response to, and representation of, the Irish Famine experience. This is echoed in quite a different way and symbolic form at other spaces of Irish Famine heritage, like, for example, the memorial at Hyde Park Barracks in Sydney or Ireland Park in downtown Toronto. Together, these spaces of cultural heritage undermine the association of monuments with straightforward narratives of the Famine experience. Instead, they present a more complex narrative of Famine memory, as well as gesturing towards more contemporary issues of world hunger. On another level they work to underscore the powerful and connective role that heritage has to play among emigrant communities. As Aplin argues:

> We need connections with both place and time to locate our present lives geographically and historically; heritage helps in both the temporal and spatial sense. It also helps us locate ourselves socially, in the sense that it is one of the things that binds communities and nations, giving a sense of group identity to both insiders and outsiders. (2002: 16)

KEY POINTS

- Heritage refers to the use of the past in the present. At its core, heritage is highly selective, culturally constructed and whether tangible or intangible, official or unofficial, fulfils a variety of oftentimes competing functions.
- The term 'dissonance' is often used to describe the disagreement or tension that can exist between the economic, cultural and political uses of heritage.
- Heritage plays a pivotal role in shaping narratives of identity and as such overlaps with other concepts like landscape and memory. It has the power to reinforce hegemonic power and strengthen group identity, while at the same time potentially excluding those who may be less dominant.
- For emigrant communities in particular, the desire to re-imagine spaces of the homeland has engendered a variety of commemorative and heritage-based practices.

Notes

1 http://whc.unesco.org/en/about/.
2 www.unesco.org/culture/ich/index.php?pg=00002.
3 www.unesco.org/culture/ich/index.php?pg=00002.

4 www.unesco.org/culture/ich/index.php?pg=00011
5 Cited by Jane Holtz Kay in 'Hunger for Memorials: New York's Monument to the Irish Famine', www.janeholtzkay.com/Articles/hunger.html, 1 February 2008.
6 www.ny.gov/governor/press/01/march15_2_01.htm, 1 February 2008.

References

Aitchison, C., Macleod, N.E. and Shaw, S.J. (2000) *Leisure and Tourism Landscapes*. London: Routledge.

Aplin, G. (2002) *Heritage: Identification, Conservation and Management*. Oxford: Oxford University Press.

Edkins, J. (2003) *Trauma and the Memory of Politics*. Cambridge: Cambridge University Press.

Graham, B. (2002) 'Heritage as knowledge: capital or culture?', *Urban Studies*, 39(5–6): 1003–1017.

Graham, B. (2007) 'The past and present of the Great Irish Famine', *Journal of Historical Geography*, 33(1): 200–206.

Graham, B. and Howard, P. (eds) (2008) *The Ashgate Research Companion to Heritage and Identity*. Aldershot: Ashgate.

Graham, B., Ashworth, G.J. and Tunbridge, J.E. (2000) *A Geography of Heritage*. London: Arnold.

Hewison, R. (1987) *The Heritage Industry*. London: Methuen.

Johnson, N. (1996) 'Where geography and history meet: heritage tourism and the Big House in Ireland', *Annals of the Association of American Geographers*, 86(3): 551–566.

Kelleher, M. (2002) 'Commemorating the Great Irish Famine', *Textual Practice*, 16(2): 249–276.

Landzelius, M. (2003) 'Commemorative dis(re)membering: erasing heritage, spatializing disinheritance?', *Environment and Planning D: Society and Space*, 21(2): 195–121.

Lowenthal, D. (1994) 'Identity, heritage and history', in J.R. Gillis (ed.) *Commemorations: the Politics of National Identity*. Princeton, NJ: Princeton University Press, pp. 41–57.

Lowenthal, D. (1998) *The Heritage Crusade and the Spoils of History*. Cambridge: Cambridge University Press.

MacCannell, D. (1992) *Empty Meeting Grounds: the Tourist Papers*. London: Routledge.

Sack, R.D. (1992) *Place, Modernity and the Consumer's World*. Baltimore, MD: Johns Hopkins University Press.

Schouten, F.F.J. (1995) 'Heritage as historical reality', in D.T. Herbert (ed.) *Heritage, Tourism and Society*. London: Mansell, pp. 21–31.

Tunbridge, J.E. and Ashworth, G.J. (1996) *Dissonant Heritage: the Management of the Past as a Resource in Conflict*. Chichester: Wiley.

Urry, J. (1990) *The Tourist Gaze: Leisure and Travel in Contemporary Societies*. London: Sage.

Urry, J. (1995) *Consuming Places*. London: Routledge.
Whelan, Y. and Harte, L. (2007) 'Placing geography in Irish studies', in L. Harte and Y. Whelan (eds) *Ireland Beyond Boundaries: Mapping Irish Studies in the 21st Century*. London: Pluto Press, pp. 175–197.

FURTHER READING

Aplin, G. (2002) *Heritage: Identification, Conservation and Management*. Oxford: Oxford University Press.
Graham, B., Ashworth, G.J. and Tunbridge, J.E. (2000) *A Geography of Heritage*. London: Arnold.
Graham, B. and Howard, P. (eds) (2008) *The Ashgate Research Companion to Heritage and Identity*. Aldershot: Ashgate.
Tunbridge, J.E. and Ashworth, G.J. (1996) *Dissonant Heritage: the Management of the Past as a Resource in Conflict*. Chichester: Wiley.

15 PERFORMANCE, SPECTACLE AND POWER

Yvonne Whelan

Introduction: Historical Geographies of Spectacle and Performance

In interrogating the historical geographies of spectacles, parades and public performances, historical and cultural geographers have foregrounded the role of public parades as rituals of remembrance, as well as choreographed expressions of power. A range of studies have focused attention on the politics of ritual and spectacle in different international contexts (Goheen, 1993; Jarman and Bryan, 1998; Cronin and Adair, 2001). Research has shown that as 'landscape metaphors', parades have a multi-faceted impact which is mediated materially and militarily, as well as through pageantry, illuminations, fanfare, music and the skilful appropriation of aspects of the past and public memory. A recurring theme in this growing body of work on the spectacle and spatiality of performance relates to the significant role of parades in identity formation, legitimation and expression. Especially noteworthy in this regard is Kong and Yeoh's analysis of state-sponsored National Day parades in Singapore. They argue that national day parades achieve their effects by 'combining the architectural spectacularity of the past and the animated spectacularity of the moment' (1997: 220). This research shows that between 1965 and 1994 the state made strenuous efforts to invent ritual and create landscape spectacles not only to construct a sense of national identity, but also to assert its hegemonic authority. As they go on to show, however, there were also instances of resistance which expose 'the cracks that particular conjunctions open in the surveillance of the proprietary powers' (de Certeau, 1984: 37, cited in Kong and Yeoh, 1997: 217).

The historical geographer Peter Goheen's work on the social valuation of public space and spectacle in late nineteenth-century Toronto reveals that the management of public spectacle was crucial to the assertion of middle-class values. He demonstrates that the city's increasingly self-confident and influential middle class became a key player in redefining the nature and meaning of urban public space (2003). Goheen's research draws attention to the symbolic centrality of the city's 1884 semi-centennial celebrations of a large parade which 'constituted a deliberate strategy of incorporation' and was designed to afford maximum opportunity for spectators (2003: 80; 1993). His analyses reveal that public space 'is now as it has always been a space of contention. It is the visible and accessible venue wherein the public – comprising institutions and citizens acting in concert – enact rituals and make claims designed to win recognition' (1998: 479). The streets of late nineteenth-century Toronto, he reveals, were frequently the preferred locale for a wide range of public parades and processions, 'from the most dignified to the most disreputable, and from the highly organised to the spontaneous' (2003: 74). Moreover, they were often timed to attract maximum publicity and designed to incorporate many of the city's neighbourhoods in a manner that afforded maximum opportunity for spectators (2003: 80; see also Trigger, 2004).

Johnson meanwhile has examined Irish commemorative parades as examples of public spectacle in dramatic form, whereby 'collective memory is maintained as much through geographical discourses as historical ones' (2003: 57). Drawing on Barthes's discussion of the cultural meaning of spectacle, she argues that the spaces in which the Peace Day and Remembrance Day celebrations were staged in Ireland in the aftermath of World War I, and the formal iconography which surrounded them, were central to the cultivation of public memory (2003: 78). Her spatial approach is echoed in Busteed's study (2005) of two sympathy parades organized in response to the execution of the three 'Manchester martyrs' in 1867, which he reads as expressions of subaltern resistance that skilfully exploited local circumstances of time, tradition and place to counteract the hegemonic force of an authoritarian regime. By focusing on the organization, route and composition of the processions, as well as on the dress and behaviour of participants, his research demonstrates how processions express resistance and reveal local power relationships. Sugg-Ryan's analysis of the Irish Historic Pageant in New York in 1913 also points to the political subtext embedded in parades. In her exploration of the ways in which the Irish Historic Pageant represented Ireland's past, she argues that 'the pageant put on a performance of Irish heritage to show not only the continuation and resilience of Irish culture in the face of upheaval and oppression by the British, but also its superiority. [...] it made a case for Home Rule and spearheaded the Irish renaissance in America' (2004: 106).

This body of research illustrates that there are a range of attributes associated with parading and public spectacle which must be closely considered by the historical geographer. Chief among these are the processional space that is the parade route, the role of the military and the forms of theatrical display that are deployed, including the use of decorations, lighting, fireworks, music and temporary structures. It is also important to consider the ways in which aspects of the past can be drawn into public spectacles, underlining the significance of public memory in the contemporary affirmation of identity narratives. The sense of a shared heritage and history oftentimes assumes a great deal of importance; as Azaryahu observes, 'national history is a prime constituent in national identity, while a sense of a shared past is crucial for the cultural viability and social cohesiveness of both ethnic communities and nation-states' (1997: 480). The past therefore constitutes a mass of symbolic capital which can be drawn upon, often rather selectively, in order to shape place identities and underpin particular ideologies. After all, group identity, which is invariably defined in contradistinction to another group, is both sustained and legitimated with reference to a resource base of shared memories (Graham et al., 2000: 18). The dynamic relationship between history and geography is underscored in the context of public spectacles when selective re-workings of past events are interwoven into public displays which act as spatialisations of memory. With these elements in mind, this chapter reflects further on the politics of performance in the urban cultural landscape, paying particular attention to the spectacular dimensions that attended one visit by a British monarch to Ireland in 1900 in order to explore further the role of public performances in the consolidation and contestation of power in the cultural landscape.

Public Spectacles of Imperial Power

The 90-year period from 1821 to 1911 witnessed a total of nine Irish visits by British monarchs, each of which occasioned a considerable measure of fanfare and ceremonial ritual: King George IV in 1821; Queen Victoria in 1849, 1853, 1861 and 1900; King Edward VII in 1903, 1904 and 1907; and King George V in 1911 (Loughlin, 2007). The pomp and pageantry surrounding these visits bore witness to the politicization of public space through the transformation of the urban landscape. On each occasion the state authorities manipulated invented ritual and landscape spectacle in order to underline the enduring importance of the monarchy to Irish culture and identity and undermine the growing threat of separatist nationalism. The spectacular elements of these visits – the street decorations, the erection of temporary structures, the soundscape created by military bands, the fireworks, the symbolic welcoming of the monarch, the royal procession – combined to create a 'web of signification' (Ley and Olds, 1988) that was skilfully spun by the state to assert its hegemonic authority. Not all of these elements were unchanging, however, and particularly noteworthy for our purposes are the varying ways in which the authorities exploited the symbolic capital in Dublin's cultural landscape in response to a shifting Irish political landscape (Whelan, 2005).

Queen Victoria's 1900 visit brings some key aspects of this strategic appropriation of urban space into sharp focus. For this visit, the royal procession from Kingstown to the Viceregal Lodge in the Phoenix Park was carefully orchestrated in a manner that reinforces the view that 'the choice of the landscape in which to stage a spectacle is not a matter of indifference, for some sites have more significance that others. [...] Parades do not simply occupy central space, but also move through space as a means of diffusing the effects of the spectacle' (Kuper, 1972: 421). To begin with, the procession passed through Dublin's more affluent southern suburbs, where many of the street names were already redolent of the city's imperial status. When the royal cortège arrived at Leeson Street Bridge, the Queen was formally welcomed and presented with the city keys and civic sword, as had happened on previous visits. What was new on this occasion was the erection of a mock medieval castle gate in place of the more traditional triumphal arch. Designed by Thomas Drew, this painted wooden structure, 70 feet in height, was presented as a replica of the medieval Baggotrath Castle, the site where, in August 1649, Royalist forces were defeated by the Roundheads, thus enabling Cromwell's entry into Dublin. This impressive showpiece was no mere decorative pastiche, therefore, but a calculated attempt to exploit the residual symbolic capital in Dublin's cultural landscape by invoking the memory of a time when Irish and Royalist forces stood together as allies.

After the presentation of the address of welcome, the royal cortège entered the city boundary and passed along Merrion Square to College Green, where it received a resounding welcome from the assembled students and fellows of Trinity College. An *Irish Times* reporter succinctly underlined the suitability of this part of the city as the setting for a dramatization of imperial power and noted the use of evanescent display as a means of interpellating the local populace:

> The scene in College Green was one that will linger long and pleasantly in the memory of all who were fortunate enough to witness it. College Green, the heart of the city, and which has from time immemorial been the central point of all processions and pageants occurring within the civic bounds, was yesterday the great central rallying point in the triumphal progress of the Queen. [...] The immense open space, the magnificent architectural surroundings – Trinity College, the Bank of Ireland, and the other splendid buildings – all helped to enhance the picturesque character of the scene. (*Irish Times*, 5 April 1900)

The procession then moved north via Dame Street and Parliament Street where it crossed the river Liffey and made its way along the quays to the Phoenix Park, bypassing completely the largely slum-ridden and poverty-stricken north inner city.

Unlike other truly imperial cities, however, where such choreographed spectacles usually proceeded smoothly, turn-of-the-century Dublin was caught in a schizophrenic position which these royal events crystallized anew. While the unionist press delighted in the city's adornment, nationalist commentators were quick to register local objections to the royal visit, many of which centred on the orchestrated politicization of social space. For example, an *Irish People* commentator protested that the 'gaudy' 'foreign' colours 'do not suggest an Irish welcome. They only indicate that wealthy Dublin Unionists have [...] given the Southern side of Dublin the appearance of an English city for the time being'. Combining anti-imperial animus with socialist critique, the writer went on to complain that every Union flag is

> an outward and visible token of the degradation of the capital – a symbol of the still dominant power of the English ascendancy. [...] Under all the banners and floral devices and glaring illuminations, was still poor old hum-drum Dublin, with its 200,000 dwellers in squalid tenements, its ruined trade, its teeming workhouses, its hopeless poverty – the once-proud capital of a free nation degraded to the level of the biggest city of a decaying and fettered province. (*Irish People*, 7 April 1900)

Capitalizing on such dissent, militant nationalists actively contested the politicization of public space by organizing a counter-demonstration in the form of a torchlight procession on the night of the royal party's arrival in Ireland. This oppositional activism reached its height some weeks later when, on 1 July, an 'Irish patriotic children's treat' was staged by Maud Gonne and Inghinidhe na hÉireann (Daughters of Ireland) as a calculated counterweight to the two 'children's treats' hosted by the Queen during her visit. This pointedly counter-hegemonic gesture proved to be the most celebrated anti-royalist event of the year (Condon, 2000: 173). Just as public space had earlier been manipulated by the state to reinforce Irish imperial loyalties, nationalists used space and spectacle to articulate Ireland's resistance to empire, thereby exemplifying the argument that 'the mimicry of the postcolonial subject is [...] always potentially destabilizing to colonial discourse, and locates an area of considerable political and cultural uncertainty in the structure of imperial dominance' (Ashcroft et al., 2002: 142).

Conclusion

This brief reading of Queen Victoria's visit to Ireland in 1900 seeks to draw attention to the material role of landscape, social space and spectacular display in the mediation of state power. Building on a burgeoning body of literature in historical and cultural geography which interrogates the politics of public spectacle in various historical contexts, it shows that as evanescent spectacles, such state occasions sought to leave an indelible mark upon the popular consciousness by exploiting both the enchanting power of grand display and the latent symbolic capital of the cultural landscape. The pomp and ceremony, the elements of display and theatricality, the emphasis on the visual and the aural, all combined in 1900 to create a 'web of signification' that was skilfully spun by the state and supported by loyal institutions. The symbolic welcome, the ceremonial address, the erection of temporary structures, the soundscape created by the military bands, the decorations, fireworks and fanfare all combined to create an imperial veneer that temporarily transformed the city into a theatre of pomp, while the royal procession itself allowed for the symbolic capture of urban space in a 'spectacularly' potent manner that reinforces Yeoh and Kong's thesis that 'the materiality of landscape does not simply provide a passive backcloth for the enactment of spectacle but its architecture and aesthetics are designed to invade the private realm and invite visual consumption of inscribed meanings' (Kong and Yeoh, 1997: 220). It is equally significant, however, that the explicitly spatial performance of state power that was evident in Dublin in 1900 was also hotly contested by the political opponents of empire who themselves used social space to stage symbolic gestures of resistance. As Ireland inched closer to Home Rule, nationalist opposition to royal tours hardened, so that while each monarch's visit to Dublin succeeded in generating widespread fervour, they were less successful in cultivating long-term imperial loyalty. In fact, the British state's efforts to invent ritual and create landscape spectacle as a means of asserting its hegemonic authority increasingly served to expose the very fault-lines it sought to conceal. By effectively galvanizing nationalist groups into oppositional activity, therefore, these spectacular events confirm the view that 'while ideology is dominant, it is also contradictory, fragmentary and inconsistent and does not necessarily or inevitably blindfold the "interpellated" subject to a perception of its operations' (Ashcroft et al., 2002: 222).

KEY POINTS

- In interrogating the historical geographies of spectacles, parades and public performances, historical and cultural geographers have highlighted the role of public parades as spectacular rituals of remembrance, as well as choreographed expressions of power.
- Research has shown that as 'landscape metaphors', parades have a multi-faceted impact which is mediated materially and militarily, as well as through pageantry, illuminations, decorations, fanfare and music and the skilful appropriation of aspects of the past and public memory.
- The dynamic relationship between history and geography is underscored in the context of public spectacles when selective re-workings

of past events are interwoven into public displays to act as a spatialization of memory and affirm narratives of identity.

- The case study of Queen Victoria's last visit to Ireland in 1900 reveals that such state occasions sought to leave an indelible mark upon the popular consciousness by exploiting both the power of spectacular display and the latent symbolic capital of the cultural landscape. It also galvanized nationalists' oppositional activity, however, who themselves used social space to stage symbolic gestures of resistance.

References

Ashcroft, B., Griffiths, G. and Tiffin, H. (2002 [1998]) *Key Concepts in Post-Colonial Studies*. London: Routledge.

Azaryahu, M. (1997) 'German reunification and the power of street names', *Political Geography*, 16(6): 479–493.

Busteed, M. (2005) 'Parading the green: procession as subaltern resistance in Manchester in 1867', *Political Geography*, 24(8): 903–933.

Condon, J. (2000) 'The patriotic children's treat: Irish nationalism and children's culture at the twilight of empire', *Irish Studies Review*, 8(2): 167–178.

Cronin, M. and Adair, D. (2001) *The Wearing of the Green: A History of St. Patrick's Day*. London: Routledge.

de Certeau, M. (1984) *The Practice of Everyday Life*. Berkeley: University of California Press.

Goheen, P. (1993) 'Parading: a lively tradition in early Victorian Toronto', in A.R.H. Baker and B. Gideon (eds) *Ideology and Landscape in Historical Perspective*. Cambridge: Cambridge University Press, pp. 330–351.

Goheen, P. (1998) 'The public sphere and the geography of the modern city', *Progress in Human Geography*, 22(4): 479–496.

Goheen, P. (2003) 'The assertion of middle-class claims to public space in late Victorian Toronto', *Journal of Historical Geography*, 29(1): 73–92.

Graham, B., Ashworth, G.J. and Tunbridge, J.E. (2000) *A Geography of Heritage*. London: Arnold.

Jarman, N. and Bryan, D. (1998) *From Riots to Rights: Nationalist Parades in the North of Ireland*. University of Ulster: Centre for the Study of Conflict.

Johnson, N. (2003) *Ireland, the Great War and the Geography of Remembrance*. Cambridge: Cambridge University Press.

Kong, L. and Yeoh, B. (1997) 'The construction of national identity through the production of ritual and spectacle', *Political Geography*, 16(3): 213–239.

Kuper, H. (1972) 'The language of sites in the politics of space', *American Anthropologist*, 74: 411–425.

Ley, D. and Olds, K. (1988) 'Landscape as spectacle: world's fairs and the culture of heroic consumption', *Environment and Planning D: Society and Space*, 6(2): 191–212.

Loughlin, J. (2007) *The British Monarchy and Ireland*. Cambridge: Cambridge University Press.

Sugg-Ryan, D. (2004) 'Performing heritage: the Irish Historical Pageant, New York, 1913', in M. McCarthy (ed.) *Ireland's Heritages*. Aldershot: Ashgate, pp. 105–122.
Trigger, R. (2004) 'Irish politics on parade: the clergy, national societies, and St. Patrick's Day processions in nineteenth-century Montreal and Toronto', *Histoire Sociale – Social History*, 37(74): 159–199.
Whelan, Y. (2005) 'Procession and protest: the visit of Queen Victoria to Ireland, 1900', in L. Proudfoot and M. Roche (eds) *(Dis)Placing Empire: Renegotiating British Colonial Geographies*. Aldershot: Ashgate, pp. 99–116.

FURTHER READING

Busteed, M. (2005) 'Parading the green: procession as subaltern resistance in Manchester in 1867', *Political Geography*, 24(8): 903–933.
Goheen, P. (1993) 'Parading: a lively tradition in early Victorian Toronto', in A.R.H. Baker and B. Gideon (eds) *Ideology and Landscape in Historical Perspective*. Cambridge: Cambridge University Press, pp. 330–351.
Kong, L. and Yeoh, B. (1997) 'The construction of national identity through the production of ritual and spectacle', *Political Geography*, 16(3): 213–239.
Ley, D. and Olds, K. (1988) 'Landscape as spectacle: world's fairs and the culture of heroic consumption', *Environment and Planning D: Society and Space*, 6(2): 191–212.

Section 6
Modernity and Modernization

16 CAPITALISM AND INDUSTRIALIZATION

Ulf Strohmayer

Introduction

Historical thinking has arguably been obsessed with epochal borderlines for most of its long existence. Often demarcated with the help of singular events – battles, royal successions or the date marking a particular technological invention has first seen the light of day – such temporal markers help to structure history and thereby render it accessible. Historical geography is no exception; in fact, its practices are often dependent on the work of historians, whose legacy paves the way for fruitful geographical encounters and scholarship. But historical accuracy is not the same as accuracy in the natural sciences; granted, we may be able – courtesy of archival evidence – to pinpoint dates attaching to events with reasonable degrees of precision, but no such act will automatically imply or lead to the construction of a context within which any such event will become meaningful.

Dating Historical Geographies

By way of an example that has a direct bearing on the topic of this chapter, let us take the year 1765 and let us focus on the invention of the modern steam engine by James Watt. Immediately, we may acknowledge that the date – '1765' – is but one possible date in sequence of possible dates: Watt himself conceived of the idea for his machine in 1763 and the actual production of the machine took place from 1775 onwards. Furthermore, Watt's machine itself was built in a line of already established steam engines, dating back at least to the late seventeenth century.

Even more crucial than these attempts at establishing historical accuracy is the importance attributed to the steam engine in and for the rise of both industrialization and capitalism, which in turn are often held to be concomitant or even synonymous with modernity. In this line of reasoning, a particular technological invention is causally linked with transformative processes in societies at large; its geography is furthermore directly linkable with both the spread of the invention (now often re-christened 'innovation' by geographers) itself and the distribution and transport of the resources required to make it work. And yet none of the above scholarly activities will be sufficient for the construction of a convincing, singular explanatory link between an innovation, in our case the steam engine, and a wider context, in our case industrialization, capitalism and/or modernity. The question we ought to ask is whether a particular invention sufficiently explains a particular process that ensues.

In the absence of clarity relating to such contextual processes, individual contributing factors (like innovations or other events) lack the orientation required for them to be(come) meaningful. And yet (again), defining wider contexts without recourse to pivotal events often becomes a fruitless task; we are, it would seem, in the company of the timeworn 'hen-and-egg' problem. More accurately perhaps to conceive of the problem in hand as one akin to the context encountered by the passionate jigsaw puzzlers of the world where a larger picture emerges from individual pieces, which in turn require a preordained place for them to 'fit'.

Returning to this chapter, the task is thus to conceptualize a link between what we seek to explain – industrialization and capitalism – and the factors that contribute to it. And just as a jigsaw puzzle requires a designer of some kind, many scholars have argued over both the ingredients and the larger picture, as they arrange to define the terms we should employ better to understand our contemporary world. Geographers in general, and historical geographers in particular, have contributed to these debates. It is to some of these that we will now turn our attention.

A prudent starting assumption would refuse to conflate the two terms that combine to form the heading to this chapter and insist that capitalism and industrialization are synonymous neither in what they describe nor in their temporal extension. Although often conflated or used to describe one another, capitalism and industrialization are two interrelated but independent historical processes. Of the two, capitalism is arguably the older and more drawn-out of the two, while industrialization can best be described as a key accelerating moment within capitalism. The notion of 'industrial capitalism' as a distinct phase within capitalism perhaps exemplifies and backs up this argument.

So, then, what precisely is capitalism and when did 'it' start? Strictly speaking, no consensus exists within historical geography and even as prominent a scholar as Robin Butlin in his seminal *Historical Geography* merely alludes to 'increasing capitalist relations of production' (1993: 148), without adding dates or any sign of further historical specificity. This vagueness is characteristic of the problem and will thus not be resolved easily.

Spatializing Capitalism

As the work of Immanuel Wallerstein has demonstrated in his influential 'world-systems analysis', capitalism can best be understood as a global process of 'geotemporal' restructuring (Wallerstein, 1988) that builds on 'the commodification of everything' (Wallerstein, 1983, 16) in the interest of generating more capital; in this definition, the 'self-expansion' of capitalism is a key element of its systemic nature. What is more, this 'growth' (or 'development') takes place on two distinct scales. First, it attaches to an increase in numerical terms, in which a localizable interaction between the costs of production, labour and exchange combines to yield surplus value, which can be re-invested into regional economies. Second, it describes a geographical process that relates spaces in an uneven manner across time, thereby creating distinct and historically contingent 'core' and periphery' regions. Starting with the expansion of mainly European trade routes in the wake of explorers like Columbus and (later) Cook and the gradual incorporation of spaces on a host of scales into a system of uneven trade and commerce, capitalism is thus originary 'global' in orientation and practice (Ogborn, 2000: 43) and has historically intimate links with processes of colonialism and imperialism discussed elsewhere in this book (Corbridge, 1987). Central to the ability of capital successfully to engage with space in this manner were three key changes to the mechanisms involved in the creations of wealth: (1) an increase in mobility across a range of contexts, ranging from modes of transportation to the dissemination of knowledges in written form (Ogborn, 2002) to the mobilization of bodies across a range of scales; (2) the development of novel forms of measuring time other than through naturally pre-given, seasonal rhythms (Ingham,

1999); and (3) the release of transfer modalities from local anchorage through the invention of novel forms of credit, paper money and internationally operating stock markets (Black, 1995; Bonney, 2001).

The precise onset of the industrialization has also been the subject of some debate (see Butlin 1993: 233). While most scholars agree that Britain set the early pace in the industrialization of its economic practices, chief amongst which was the move towards manufacturing as the prime source of societal wealth generation, attaching a more precise temporal and geographical frame to this process – often referred to as the 'Industrial Revolution' – has proven to be wrought with difficulties (Pinard, 1988; Royle, 1991; Stobbard, 1996; Townshend, 2006). What Butlin and others have referred to as the 'proto-industrialisation' of rural landscapes exhibiting (Butlin, 1993, esp. 227–220; Cohen, 1990) tentative signs of industrialization, combined with the fact that not all nations or regions experienced the process during comparable temporal intervals or indeed at the same time, renders any singular history of industrialization impossible while making its historical geography imperative.

Most authors agree on the accelerating potential that processes of industrialization offered and continue to offer to social systems organized along capitalist principles. Key to this was not just the increase in productive output offered by new technologies but a more general amplification of the speed at which societies operated. The spread of information through telegraphic wires during the second half of the nineteenth century, as well as the increase in transportation speeds through the spread of railroads at the same time, contributed centrally to the ability of capital to increase its rate of circulation and thereby to proliferate. A different way of looking at this process would be to analyse it through the lenses of mobility, stressing the different forms of mobilization that were conditions of possibility of capital to operate across scales. From the mobility of capital in the form of credit and investment to the mobility of resources and labour (Cawley 1980; Bourdelais 1984), capitalism, as Neil Brenner has argued, is historically characterized by its ability to exploit differences between spatial fixity and flow across scales in ongoing processes of territorialization and re-territorialization (1998). And note again how these processes involve *possible* mobility traverse across scales: the mobility of labour, for instance, involves daily commuting patterns, nineteenth-century mass-migration to cities and present-day migration across continents and digital forms of mobility, linking homework support in India with the USA.

Such processes characterizing capitalist modes of production, as Marx famously referred to them, cannot ever be stable. Quite the contrary: instability is the engine of capitalist accumulation and the recognition and subsequent acting on differentials (of prices, wages, amount of influence, information, etc.) effectively forms the accelerator; the resulting system has thus historically operated through a variety of man-made crises (Black 1989), which have arguably become a distinguishing hallmark of modern, globalized societies ever since. If capitalism qua industrialization has created industrialized landscapes – Yorkshire, the Ruhr, Northern France are perhaps the best-known examples of such altered spaces – we ought to remember that most of these landscapes have undergone further transformative processes since World War II in the context of the de-industrialization of what came to be known as 'old' or 'traditional' industries (Linehan, 2000). The flexibility of the (capitalist) system thus periodically becomes a crisis somewhere: capitalism is characterized precisely by its ability to produce and reproduce spaces through the ability and willingness of capital to exploit price differentials on a host of scales (Wallerstein, 1983; Steinberg, 2000) at the 'price' of abandoning entire regions or nations. Not coincidentally, the notion of 'crisis' (and perhaps even more of 'abandon') also attaches to the incorporation of 'nature' into the capitalist system: conceptualized as a resource only, and thus measured as a direct costs only, nature continues to be accorded a status similar to the one accorded to labour in that it registers only when acknowledged to be scarce. Hence the global operations of capitalism have proven to be globally destructive vis-à-vis nature, as Moore has demonstrated with regard to the industrialization of the production of sugar in early modern capitalism (Moore, 2000; see also D'Souza, 2006).

Capitalism and the City

Implicated in, as well as driving, this process were processes of urbanization that went hand in hand with both capitalism and modernization – and continue to do so (see also Chapter 11). In this context, it is important to differentiate between the role of cities in the rise of capitalism, that is the development of increasingly globally networked centres of commerce, capital and transport, and the capitalization of the built environment that turned cities themselves into an operating field of global capital. The former process accompanied the rise of capitalism from its inception while the latter process is characteristic mostly of the nineteenth-century exponential growth that took place in many cities (Harvey, 1989). Common to both and intimately related to the development of capitalism are the development of spatial divisions of labour (Moore, 2002), as well the regulation of spaces from the local scale upwards to allow the private interests of merchant and industrial capital to beget commonly shared and publicly operating practices (Harreld, 2003).

Caught up in the emergence and continuous reorganization of public and private spaces is the transformation of gender relations and the re-inscription of landscapes of work, home and leisure that are formed in and in turn support this process (Horrell and Humphries, 1995). The process referred to as 'housewifization' by Maria Mies (1989, esp. 104–119), for instance, is clearly implicated not just in the development but furthermore in the support and maintenance of capitalism as a socio-cultural process. Part and parcel of such gendered processes were changes that occurred in the nineteenth century, which arguably changed the nature of capitalism as much as it did change the gendering through which it worked. Typically, capitalism is conceptualized as a mode of production and our discussion to date has attempted to highlight key aspects contributing to changes in the production process and context during capitalism.

The move towards consumption as arguably the key engine of capitalist development in Western societies – coinciding with what Marx famously called the transformation of mere 'use value' into a commodity that is from hence primarily characterized by an independent exchange value. Increasingly, this latter value is constituted through recourse to visually engineered forms of display in the form of 'spectacles' (Pinder, 2000). Ultimately, the process involves the elevation of leisure in its many forms and guises to a central tenet of capitalism: from tourism (Gilbert, 1999; Gilbert and Hancock, 2006) to entirely visual forms of entertainment like panoramas, dioramas and the cinema (Clarke and Doel, 2005) and in the process 'corrupting' older basic economic practices (Strohmayer, 1997).

Conclusion

Ultimately, the adaptability of capitalism to changing circumstances and technologies has demonstrated itself through recent developments, chief amongst which must count the rise of the network society in what has been referred to as the 'second industrial revolution'. It is well worth noting, even in a book primarily centred on knowledge and practices in historical geography, how present-day developments in the context of industrialization and modernity continue processes of individualization that started during the Enlightenment some 250 ago only to complement them with a notion of 'competitiveness' that has become all-pervasive in recent years. From a more narrowly construed historical viewpoint, it is worth noting that the speed of development has again been intensified during this ongoing process: while industrialization took two centuries to reach all parts of the world, the information revolution took two decades to reach most (Castells, 1996), impacting further on the flexibility of capital to exploit differentials and thereby to control space through the constant manipulation of work practices (including wages, work permits, working conditions, processes of de-skilling, etc.) on a global scale.

KEY POINTS

- Capitalism characterizes a drawn-out process of social change that relies centrally on the production of commodities and the development of systems of commodity exchange.
- It is important to contextualize the rise of capitalism within the intimately related histories of exploration, colonialism, imperialism and industrialization. As such, it is imperative to conceptualize capitalism as having been a *global* system even before the currently prevailing notion of globalization took roots.
- The relationship between changes in the mode of production such as capitalism to technological change is a multi-faceted one that requires geographical sensitivity, alongside an awareness of larger social and economic change.
- An integral part of capitalism has always been a continuous process of de- and re-territorialization of the space.

References

Black, I. (1989) 'Geography, political economy and the circulation of finance capital in early industrial England,' *Journal of Historical Geography*, 15(4): 366–384.

Black, I. (1995) 'Money, information and space: banking in early-nineteenth century England and Wales', *Journal of Historical Geography*, 21(4): 398–412.

Bonney, R. (2001) 'France and the first European paper money experiment', *French History*, 15(3): 254–272.

Bourdelais, P. (1984) 'L'industrialisation et ses mobilités', *Annales: Economies, Societés et Civilisations*, 39(5): 1009–1019.

Brenner, N. (1998) 'Between fixity and motion: accumulation, territorial organization, and the historical geography of spatial scales', in *Environment and Planning D: Society and Space*, 16(4): 459–481.

Butlin, R.A. (1993) *Historical Geography: Through the Gates of Space and Time*. London: Edward Arnold.

Castells, M. (1996) *The Rise of the Network Society*. Oxford: Blackwell.

Cawley, M. (1980) 'Aspects of rural–urban integration in western Ireland', *Irish Geography*, 13: 20–32.

Clarke, D.B. and Doel, M. (2005) 'Engineering space and time: moving pictures and motionless trips', *Journal of Historical Geography*, 31(1): 41–60.

Cohen, M. (1990) 'Peasant differentiation and proto-industrialisation in the Ulster countryside: Tullylish 1690–1825', *Journal of Peasant Studies*, 17(3): 413–432.

Corbridge, S. (1987) 'Industrialisation, internal colonialism and ethnoregionalism: the Jharkhand, India, 1880–1980', *Journal of Historical Geography*, 13(3): 249–266.

D'Souza, R. (2006) *Drowned and Dammed: Colonial Capitalism and Flood Control in Eastern India*. Oxford: Oxford University Press.

Gilbert, D. (1999) '"London in all its glory-or how to enjoy London": guidebook representations of imperial London', *Journal of Historical Geography*, 25(3): 279–297.

Gilbert, D. and Hancock, C. (2006) 'New York City and the transatlantic imagination: French and English tourism and the spectacle of the modern metropolis, 1893–1939', *Journal of Urban History*, 33(1): 77–107.

Harreld, D. J. (2003) 'Trading places: the public and private spaces of merchants in sixteenth-century Antwerp', *Urban History*, 29(6): 657–669.

Harvey, D. (1989) *The Urban Experience*. Baltimore, MD: Johns Hopkins University Press.

Horrell, S. and Humphries, J. (1995) 'Women's labour force participation and the transition of the male-breadwinner family, 1790–1865', *Economic History Review*, 48(1): 89–117.

Ingham, G. (1999) 'Capitalism, money and banking: a critique of recent historical sociology', *British Journal of Sociology*, 50(1): 76–96.

Linehan, D. (2000) 'An archaeology of dereliction: poetics and policy in the governing of depressed industrial districts in interwar England and Wales', *Journal of Historical Geography*, 26(1): 99–113.

Mies, M. (1986) *Patriarchy and Accumulation on a World Scale*. London: Zed Books.

Moore, J. W. (2000) 'Sugar and the expansion of the early modern world-economy: commodity frontiers, ecological transformation, and industrialization', *Review*, 23(3): 409–433.

Moore, J. W. (2002) 'Remaking work, remaking space: spaces of production and accumulation on the reconstruction of American capitalism, 1865–1929', *Antipode*, 34(2): 176–204.

Ogborn, M. (2000) 'Historical geographies of globalisation, c. 1500–1800', *Modern Historical Geographies* (eds B. Graham and C. Nasg). Harlow: Longman, pp. 43–69.

Ogborn, M. (2002) 'Writing travels: power, knowledge and ritual on the English East India Company's early voyages', *Transactions of the Institute of British Geographers*, 27: 155–171.

Pinard, J. (1988) 'The impact of industrialization on the development of the rural environment of the Poitou-Charentes district in the 19th and 20th centuries', *Geografiska Annaler B*, 70(1): 219–225.

Pinder, D. (2000) '"Old Paris is no more": geographies of spectacle and anti-spectacle', *Antipode*, 32(4): 357–386.

Royle, S. (1991) 'The socio-spatial structure of Belfast in 1837: evidence from the First Valuation', *Irish Geography*, 24(1): 1–9.

Steinberg, P. (2000) 'Place, power, and paternalism: imagined histories and welfare capitalism in Burrillville, Rhode Island, 1912 to 1951', *Urban Geography*, 21(3): 237–260.

Stobbard, J. (1996) 'Geography and industrialisation: the space economy of north-western England, 1701–1760', *Transactions of the Institute of British Geographers*, 21(4): 681–696.

Townshend, C. (2006) 'County versus region? Migrational connections in the East Midlands, 1700–183', *Journal of Historical Geography*, 32(2): 291–312.

Wallerstein, I. (1983) *Historical Capitalism*. London, Verso.

Wallerstein, I. (1988) 'The inventions of TimeSpace realities: towards an understanding of our historical systems', *Geography*, 73(4): 289–297.

FURTHER READING

Black, I. (1995) 'Money, information and space: banking in early-nineteenth century England and Wales', *Journal of Historical Geography*, 21(4): 398–412.

Brenner, N. (1998) 'Between fixity and motion: accumulation, territorial organization, and the historical geography of spatial scales', *Environment and Planning D: Society and Space*, 16(4): 459–481.

Moore, J.W. (2000) 'Sugar and the expansion of the early modern world-economy: commodity frontiers, ecological transformation, and industrialization', *Review*, 23(3): 409–433.

Strohmayer, U. (1997) 'Technology, modernity, and the restructuring of the present in historical geographies', *Geografiska Annaler*, 79B(3): 155–170.

17 CULTURES OF SCIENCE AND TECHNOLOGY

Ulf Strohmayer

Introduction

A dimension shared by many definitions of 'modernity' is the direct involvement of both 'science' and 'technology' in the construction of modern objects and subject alike. Going hand in hand with the notion of 'progress', technologically infused and scientifically backed modernity emerges as a categorically different and at one stage 'new' context in which humankind operates. The preceding chapter looked more closely at the steam engine as one such context-changing object and analysed its importance within the ongoing history of capitalism. This chapter will analyse the changing cultures attaching to both modern science and technology on a broader canvas.

Customarily, the idea motivating such and related definitions of modernity is that of a categorical break: the 'modern' world is thought to be substantially different from preceding worlds so that any attempt at describing it adequately requires a noun ('modernity') or an adjective ('modern') for it to succeed. In other words, the argument is that not only did the world change, but it changed substantially, irreversibly even. Until fairly recently, 'modernity' was thought to characterize an open-ended process; presently, much debate centres around proposed end-points demarcating this process in the form of 'post-modern' or 'hyper-modern' characteristics. We shall investigate these claims further on in this chapter; for now, let us focus on the historical geographies attaching to their objects and practices. The first thing to note is that most writings in historical geography did not conceptualize either science or technologies as separate fields of enquiry until the late 1980s when a string of key papers and books made the geographies that underpin scientific progress and technological advances a key focus of individual and collective endeavours (Livingstone, 1990). The motivation for these publications was closely associated with the 'spatial turn' across the human sciences and argued convincingly that practices emanating from both science and technologies were intimately tied up with their spaces of production. As concerns the historical geography of technologies, this argument was strictly speaking not a new one, especially not in a discipline with a rich tapestry of scholarship relating to questions of innovation; the questions attaching to a historical geography of science were more problematic. After all, as Shapin has mentioned, did not science purport to be a universal practice that was expressly not tied to specific geographies (1998)? It is perhaps one of the more original insights that have emanated from within historical geography and related areas of academic interest during the last two decades to argue forcefully that while scientific claims to truth are based on a universal language of non-place specific evidence, the making of science emanated from 'a variety of practices whose conceptual identities were the outcomes of local patterns of training and socialisation' (Shapin, 1998: 6). It is to these contextual forces that we shall now turn our attention.

Modern Science

The question when to date the arrival of modern scientific manners of thinking is obviously tied in with a broader conceptualization of modernity and modern practices. Just as obviously, it is connected to the emergence of particular technologies and technologically infused practices. For instance, the foundational importance customarily attributed to the telescope as part of Galileo Galilei's discovery of the moons orbiting Jupiter in 1610 – a discovery that constituted a contradiction of the then dominant Aristotelian astronomy – clearly points towards a close connection between technology and science. Galileo's life and scientific thinking formed part of the so-called modern 'scientific revolution' stretching across the seventeenth century, beginning with the work of Galileo and Descartes and leading straight towards broader changes in society during the Enlightenment that followed (see Livingstone and Withers, 2005 for a critical interpretation of this and other 'revolutions'). Postulating the seventeenth century as a watershed and given the predominance accorded to 'scientific' manners of thinking since then comparatively denies the validity of previously existing practices both within Europe and in other parts of the world. To some extent, this is intentional and has become a key moment in the foundational practice to elevate a 'Eurocentric' reading of science and technology to a predominant and all but exclusive status over and against parallel or competing scientific practices. However, this 'heroic' story of reason gradually conquering novel ground masks the absence of a categorical dividing line between 'magical' (i.e. 'non-scientific') and 'rational' modes of thinking in the sixteenth, seventeenth and eighteenth European centuries, as Livingstone has argued (1990). Also, the contributions made by China or Arabic countries to the formation of cultures of science and scientific practices often receive but scant recognition in the history of science. The reason for this is largely due to a teleological impulse that dominates much writing of history: the 'reading backward' of historical changes and developments as if they could only have led to a future present. Thus the centrality of 'science' to the development of European modes of existence and to the advance of Western economies and cultures to a recent position vying for global supremacy often led to the presupposition that the predominant sites of scientific progress were located in Europe and, from the nineteenth century onwards, in North America.

Most of the ensuing scholarship took scientific progress to be intimately tied in with the Enlightenment and its open emphasis on evidence-based forms of understanding, as well as its often-implied turn towards non-metaphysical knowledge based on anti-authoritarian forms of experience (Livingstone, 1990; Livingstone and Withers, 2005). Coinciding for many with the advance of modernity, the Enlightenment in turn was seen to build on the opening towards new forms of geographic knowledge and practices that dominated a preceding Age of Exploration and Discovery. In fact, many commentators point towards the Portuguese Prince Henry the Navigator and his voyages of discovery as the founding moment of the scientific, evidence-based rupture that was to beget the scientific edge of Europe over and against spaces embodying competing notions of science and progress (Livingstone, 1990: 364; see also Law, 1987). Science, in other words, contributed greatly to the geographical feasibility of the European conquests since the days of Columbus (Hugill, 1995), as it did play a major part in the subjugation of nature in the context of these often-invasive new practices (Wesseling, 1995). In this highly geographical story, the 'use-value' of science to the construction, expansion and maintenance of economic domination and (later) empire is stressed at the expense of a more inward orientated, curiosity-driven and thus alternative history of science. In addition, as Shaw has argued, the use especially of new scientific tools in the construction of 'modern' maps was influenced both by the needs of capitalism to locate resources, as it was required for the construction of the modern nation-state to demarcate and control its territorial extension (Shaw, 2005). If this impact of pre-colonial and colonial travels and their associated practices in the creation and dissemination of economically viable knowledge is perhaps a tad obvious – this, after all, forms a formally undisputed core in the history of geographic practices – recent geographical writing has turned to the narratives that made the highly embodied act of travelling itself a central focus of an ensuing narrative (Naylor, 2005).

More recent scholarship in historical geography, following the pioneering work of Steven Shapin (Shapin and Schaffer, 1985) and Bruno Latour (1988), have added another geography of science to the story just told. According to their work, the development of novel practices in the realm of science – from medical to agricultural sciences – depended directly on networked forms of information: as David Livingstone has argued (2005; see also Ogborn, 2004; Driver, 2004 stresses the importance of 'logbooks' in the construction of knowledge of far-away places), science, writing and technologies of dissemination must be thought as intimately related. Crucially, the moment we allow for such networks to become part and parcel of a history of scientific progress, geographical differences become an integral part of this history, affecting both writers and audiences equally. And note again that such networks could be effective in the form of realized networks (Finnegan, 2004) and through 'imagined' intellectual dialogues (Mayhew, 2005).

No less crucial were the means through which trust could be established across space. Disseminating knowledge in an age dominated by distance, shorter life spans and the absence of immediate forms of correspondence (albeit one in which Latin still provided a lingua franca amongst those participating in scientific discourses), required modes of standardizing scientific experiences. Crucially involved in this process were key sites in the production of knowledge such as the laboratory, the museum, the archive, the field and the garden (Livingstone, 2000; Withers, 2002), publishing houses (Withers, 2001) or the no less critical space of the 'European-style schoolroom' (Shapin, 1998: 7). Irreducible key 'sites of practice' (Withers, 2007: 63), they were centrally involved in the ordering and classification of the world throughout modernity, at least from the Enlightenment onwards (Withers, 1996). It is in these spaces that a second set of standardizing practices establishes their work: techniques such as the map, the thermometer, the clock or the graph all take roots within newly developing spaces and work to ensure that novel scientific insights can be compared across space (Law, 1987; Latour, 1998; Glennie and Thrift, 2005) and are thereby used in the interests and legitimacy of those imperial ambitions that were invoked further up (Mann, 2003). The notion of 'use' here, as we need to stress from the outset, was restricted by gender and class to a highly selective group within modernizing societies; in fact, as Shapin and others have argued, where contemporary scientific practices rely on 'expertise' and institutions to establish trust across space, early modern scientific practices relied on the figure of, and practices associated with, the 'gentleman scientist' as a trustworthy individual engaged in the advance of knowledge.

Modern Technology

As the history of modern science has shown, separating science from technology is ultimately a rather fruitless task. In other words, there is a clear and present link between a developing scientific culture and technology that has manifested itself in a variety of forms throughout human history: a relationship of mutual dependency and of reciprocal fertilization. Perhaps the best example of this link comes in the form of the extensive 'annexe' folios published in conjunction with Diderot and d'Alembert's *Encyclopédie* from the middle of the eighteenth century onwards. A key moment in the European Enlightenment, these depict technological objects with the same loving embrace of detail and accuracy normally reserved for more immediately aesthetically pleasing, natural objects. Arguably, what is expressed in the *Encyclopédie* and other modern publications is none other than a peculiarly 'modern' infatuation with technologies and technological progress, both of which came to characterize 'modernity' more than perhaps anything else. The telescope has been mentioned already; placed alongside such inventions as the printing press (Eisenstein, 1979), modern tools, the train (Schivelbusch, 1987) and the steam boat, the telegraph, the electric generator and photography (to name but a few), we see a highly familiar picture emerging that is populated with expressions of technological advances. It is crucial to note, however, that these modern objects were instrumental not only for the advancement of science (see Ryan, 2005 for a fascinating account of the use of photography in the context of nineteenth-century geographical knowledge) but often created entirely novel cultures and expectations. In fact, we could go as far as asserting that technological innovations only become new technologies if and where they assume the mantle of innovations that are capable of 'shocking' established practices to the point of replacing them with new modes of doing, organizing, distributing and supervising. Hägerstrand's pioneering and hugely influential work on the diffusion of technological innovations and ensuing practices, although not strictly speaking work that classifies as 'historical', serves as a timely reminder of the centrality accorded to technological change in geographical scholarship (1968).

In the context of capitalism (see the preceding chapter), such innovations become instrumental in creating new markets or of rhythmically re-animating existing markets; new technologies, we ought to remember, mostly substitute older ways of achieving a comparable result with a new, and purportedly better, manner of achieving results. Trains gradually replaced horse-drawn coaches just as the book and the newspaper gradually replaced traditions of oral story-telling and the town-crier. Across the board of technological change, there appear to be two tendencies that apply to most: an increase in speed and a move towards increasingly individualized forms of relating from one person to another (Kern, 1983). In fact, we can imagine re-telling the history of modernity in a meaningful manner by focusing solely on the ensuing – and related – processes of individualization and increases in speed. Furthermore, the differentiated nature of this experience in different places of the globe has given rise to classifications of the world into differently 'modernized' parts, of which the well-worn differentiation between 'first' and 'third' worlds is perhaps the best known.

However, nineteenth-century technological change did not merely impact on everyday life and culture in the form of increased speeds and privacy; another, no less profound change occurred in the realm of visual media through the invention of photography and film, eventually leading to a visualization of everyday practices and the proliferation of novel forms of relating to the world through visual means. The rise of advertisement, for instance, or the thorough saturation of the public sphere with visual forms of argumentation clearly has changed the way we interact with each other since the dawn of modernity. Even the internet, to leapfrog towards the contemporary world, clearly owes much of its success to its fundamentally visual nature.

Both modern science and technology arguably contributed to what David Harvey famously summarized as a compression of space and time during modernity (1990: 240; see also Sheppard, 2006; by contrast, see

Kirsch, 1995): the acceleration of experience and of economic practices. The consequences associated with the trope of 'space–time compression' within the capitalist system have been analysed in the preceding chapter; suffice it to say that the contemporary, twenty-first century has witnessed both the continuation and acceleration of the trend analysed by Harvey through telecommunication technologies like the internet, and brought about a fundamentally different but no less 'technologized' experience of the present moment. In all of this history, it is imperative to remember that such changes not only impact upon local realities in a highly differentiated manner and thus leading to 'uneven' forms of development; technological and scientific change continue to be embedded in (where they are not an outright production of) and thus contributing to specific interests of states and corporate enterprises alike. Neil Smith's (1992) analysis of the contribution of geographic information systems (GIS) to the first Iraq war can serve as a timely reminder of broader contexts that occasionally are forgotten in the annals of technological change.

KEY POINTS

- Modernity, science and technology form a key triad that underlies many analyses in historical geography. In many instances, arguments concerning these three distinct elements operate in a circular manner by employing one term ('modernity') to explain another ('science').
- Commonly accepted histories of science and technology are often implicity or explicitly Eurocentric in scope, method and orientation.
- Both science and technologies and embedded within (and thereby constituting) clearly articulated geographies that explain the concrete emergence and the sustained maintenance of key modern practices.
- Speed and individualization are two key processes that characterize the cultures associated with modern technologically infused practices.

References

Driver, F. (2004) 'Imagining the Tropics: views and visions of the tropical world', *Singapore Journal of Tropical Geography*, 25(1): 1–17.

Eisenstein, E. (1979) *The Printing Press as an Agent of Change: Communications and Cultural Transformation in Early Modern Europe* (2 volumes), Cambridge: Cambridge University Press.

Finnegan, D. (2004) 'The work of ice: glacial theory and scientific culture in early Victorian England', *British Journal for the History of Science*, 37: 29–52.

Glennie, P. and Thrift, N. (2005) 'Clocks and the temporal structures of everyday life', in D. Livingstone and C. Withers (eds) *Geography and Revolution*. Chicago: University of Chicago Press, pp. 160–198.

Hägerstrand, T. (1968) *Innovation Diffusion as a Spatial Process*. Chicago: University of Chicago Press.

Harvey, D. (1990) *The Condition of Postmodernity: An Enquiry into the Origins of Cultural Change*. Cambridge, MA: Blackwell.

Hugill, P. (1995) *Geography, Technology and Capitalism. World Trade since 1432*. Baltimore, MD: Johns Hopkins University Press.

Kern, S. (1983) *The Culture of Space and Time, 1880–1918*. Cambridge, MA: Harvard University Press.

Kirsch, S. (1995) 'The incredible shrinking world? Technology and the production of space', *Environment and Planning D: Society and Space*, 13(5): 529–555.

Latour, B. (1998) *The Pasteurization of France*. Cambridge, MA: Harvard University Press.

Law, J. (1987) 'On the social explanation of technical change: the case of Portuguese maritime expansion', *Technology and Culture*, 28: 227–252.

Livingstone, D. (1990) 'Geography, tradition and the Scientific Revolution: an interpretative essay', *Transactions of the Institute of British Geographers*, 15: 359–373.

Livingstone, D. (2000) 'Making space for science', *Erdkunde*, 54(4): 285–296.

Livingstone, D. (2005) 'Text, talk and testimony: geographical reflections on scientific habits. An afterword', *British Journal for the History of Science*, 38: 93–100.

Livingstone, D. and Withers, C. (2005) 'On geography and revolution', in D. Livingstone and C. Withers (eds) *Geography and Revolution*. Chicago: University of Chicago Press, pp. 1–21.

Mann, M. (2003) 'Mapping the country: European cartography and the cartographic construction of India, 1760–1790', *Science, Technology and Society*, 8(1): 25–46.

Mayhew, R. (2005) 'Mapping science's imagined community: geography as a Republic of Letters', *British Journal for the History of Science*, 38: 73–92.

Naylor, S. (2005) 'Historical geography: knowledge, in place and on the move', *Progress in Human Geography*, 29: 626–640.

Ogborn, M. (2004) '*Geographia*'s pen: writing, geography and the arts of commerce, 1660–1760', *Journal of Historical Geography*, 30: 294–315.

Ryan, J. (2005) 'Photography, visual revolutions, and Victorian geography', in D. Livingston and C. Withers (eds) *Geography and Revolution*. Chicago: University of Chicago Press, pp. 199–238.

Schivelbusch, W. (1987) *The Railway Journey: the Industrialisation and Perception of Time and Space*. Berkeley: University of California Press.

Shaw, D. (2005) 'Mapmaking, science and state building in Russia before Peter the Great', *Journal of Historical Geography*, 31(3): 409–429.

Shapin, S. (1998) 'Placing the view from nowhere: historical and sociological problems in the location of science', *Transactions of the Institute of British Geographers*, 23: 5–12.

Shapin, S. and Schaffer, S. (1985) *Leviathan and the Air-pump: Hobbes, Boyle, and the Experimental Life*. Princeton: Princeton University Press.

Sheppard, E. (2006) 'David Harvey and dialectical space-time', in N. Castree and D. Gregory (eds) *David Harvey: a Critical Reader*. Oxford: Blackwell, pp. 121–141.
Smith, N. (1992) 'History and philosophy of geography: real wars, theory wars', *Progress in Human Geography*, 16(2): 257–271.
Strohmayer, U. (1997) 'Technology, modernity and the re-structuring of everyday geographies', *Geografiska Annaler (Series B)*, 79(3): 155–169.
Strohmayer, U. (2007) 'Engineering vision: the Pont-Neuf in Paris and modernity', in A. Cowan and J. Steward (eds) *The City and the Senses: Urban Culture since 1500*. Basingstoke: Ashgate, pp. 75–92.
Wesseling, H. (1995) 'The expansion of Europe, the division of the world and the development of science and technology', *European Review* 3(3): 257–263.
Withers, C. (1996) 'Encyclopaedism, modernism, and the classification of geographical knowledge', *Transactions of the Institute of British Geographers*, 21(1): 275–298.
Withers. C. (2001) *Geography, Science and National Identity: Scotland since 1520*. Cambridge: Cambridge University Press.
Withers, C. (2002) 'Constructing "the geographical archive"', *Area*, 34(3): 303–311.
Withers, C. (2005) 'Geography, science and the scientific revolution', in D. Livingston and C. Withers (eds) *Geography and Revolution*. Chicago, University of Chicago Press, pp.75–105.
Withers, C. (2007) *Placing the Enlightenment: Thinking Geographically about the Age of Reason*. Chicago: University of Chicago Press.

FURTHER READING

Kirsch, S. (1995) 'The incredible shrinking world? Technology and the production of space', *Environment and Planning D: Society and Space*, 13(5): 529–555.
Livingstone, D. (1990) 'Geography, tradition and the Scientific Revolution: an interpretative essay', *Transactions of the Institute of British Geographers*, 15: 359–373.
Mayhew, R. (2005) 'Mapping science's imagined community: geography as a Republic of Letters', *British Journal for the History of Science*, 38: 73–92.
Shapin, S. and Schaffer, S. (1985) *Leviathan and the Air-pump: Hobbes, Boyle, and the Experimental Life*. Princeton, NJ: Princeton University Press.
Withers, C. (2005) 'Geography, science and the scientific revolution', in D. Livingston and C. Withers (eds) *Geography and Revolution*. Chicago: University of Chicago Press, pp. 75–105.

18 MODERNITY AND DEMOCRACY

Ulf Strohmayer

Introduction

Most of the substantive content of the preceding two chapters has often been summarized underneath the headings of 'modernity' and 'modernization'. In this common line of reasoning, the historical geographies that characterize capitalism, science and technology all appear to be connected to, if not causally explained by, processes related to the modernization of societies or aspects thereof. In fact, few scholars and commentators customarily differentiate sufficiently between any of these concepts, resulting in a situation where they are often used interchangeably. In this manner, concepts are effectively expected to shed light on each other in an all-too-often circular manner. Inspired by such linguistic transpositions, 'modernity' is thus not infrequently thought to explain 'capitalism' (or vice versa) or 'technology' is held to be responsible for the development of a 'science-based' economy. Part of the blame for such rhetorical imprecision must be shouldered by the common and deceptively unproblematic use of the adjective 'modern' in particular. As we have seen in many parts of the present volume but especially in the chapter concerned with science and technology, the word 'modern' has become a ubiquitous term characterizing many different processes, states or contexts as being either '(merely) new', 'categorically different' or 'epochal', thus confusing, rather than clarifying, attempts at creating knowledge about the world we inhabit.

Differentiating 'Modernity'

That said, the themes of 'modernity' and 'modernization' have long been defining topics within historical geography, if not always in an explicitly thematized form. Often implied in notions of 'change' or 'adaptation', the idea of 'modern' furthermore entails a more fundamental sea-change: the idea of a world that is no longer routinely comparable with what preceded it because it harbours 'modern' material items, 'modern' processes and ideas or 'modern' structures. Traditionally, the 'preceding' epoch to what historians and other human scientists have called 'the modern epoch' was summarized under the no less grand headings of the 'medieval age' or, more simply put, the 'pre-modern world' (but see Lilley, 2004 for a critique of this traditional view). In the run-up to the end of the second millennium, this was supplemented with a further demarcation in the form of a new and 'post-modern' era, seemingly putting an end to modernity. What is meant by any of these periodizations ('modernity') hinges nominally on what is meant by the adjective ('modern'); crucially, however, any attempt to answer the time-honoured question 'What does it mean to be modern?' in turn depends upon the temporal and geographical limits contextualizing the use of the term (Toulmin, 1990, esp. 5–44). The notions of 'modern', 'modernity' and 'modernization', in other words, are constructed in a circular manner requiring historical and geographical sensitivities and precision. This is precisely what historically minded scholars and historical geographers have been aiming to achieve, especially during the last 30 or so years during which historical scholarship increasingly became more aware of larger contexts, trends and trans-cultural forces (Ogborn, 1998; Tang, 2008).

Broadly speaking, scholars focus on three aspects of social and cultural change when they invoke or otherwise use the term 'modernity'. First, 'modernity' is often synonymous with the attempt by humankind to master nature in its many shapes and guises. Part and parcel of this definitional field, which is furthermore coloured by a wide-ranging notion of progress, are key advances in science and technology. Second, modernity designates a thoroughly geographical process of incorporation that encompasses explorations and colonialism, and which culminates in present-day processes associated with globalization. The extent to which this process is synonymous with the development and spread of capitalism is open to some debate. Third, the discursive field designated by the idiom 'modern' in its many forms furthermore characterizes a political and thoroughly normative project associated with secularization processes, democracy and justice. The first two of these foci inform various chapters in *Key Concepts in*

Historical Geography, not least the two preceding chapters in this section; it is to the last of these, the political project associated with modernity – which arguably culminates in practices associated with democracy – and the historical geographies that it begets that this chapter will now turn its attention.

In its most traditional and materially present form, the link between modernity as a political project and geography is forged through key historical sites and moments. Not surprisingly, geographers have picked up the threads woven by historians by focusing more explicit attention onto the spaces that were involved and causally connected to, amongst others, revolutions, the formation of nation-states and diplomatic negotiations (Harvey, 1979; Heffernan, 2001; see also Section 2). Surprisingly, however, the number of historical geographers who engage with such primary sites of history remains relatively small; for example, an unequivocal and comprehensive historical geography of the storming of the Bastille still awaits its author (but see Bonnemaison, 1998 for a contemporary analysis of the use of the symbol and Bassett, 2008: 905 for a theoretical invocation of the Bastille). The absence of such explicitly site-specific work should not be seen to imply that the political aspects associated with modernity have not been analysed by historical geographers. Far from it, but crucially, such explorations increasingly focus on the representations and associated practices through which such aspects or connections were forged (Konvitz, 1990; Maddrell, 1998; Helleiner, 1999; Clayton, 2000; Power, 2001; Graham and Shirlow, 2002; Moran, 2006), including studies of the construction of geographical space in the process of modernizing societies (Sparke, 1998; Collier and Inkpen, 2003). Implicitly, most of the work undertaken in this discursively cognizant manner focuses on particular aspects of modernity that imply politics on a host of scales. Here, many of the themes analysed elsewhere in this book often acquire political nuances or become 'political' as a result of being contextualized within novel theoretical frameworks which in turn invite new interpretations of modernity (Strohmayer, 1995). Related to such analyses are historical explorations that seek to locate modernity in everyday historical, material and political contexts. Exemplary here is perhaps Miles Ogborn's interpretation of institutional spaces in early modern London (1998), which refrains nonetheless from embedding politics within the spaces it analyses more forcibly.

Modernity and the Public Sphere

In addition to such non-metaphorical place-driven and space-driven analyses of the historical geographies of modernity, 'politics' enters into the analytical frame mainly through two distinct and different avenues, both of which have motivated research and thinking in historical geography. The first of these attaches to the idea of 'politics' emanating from within a sphere where people meet, exchange ideas and deliberate on collective forms of action. A second interpretation of the nexus between modernity and politics has fomented around practices associated with modern forms of government. From a geographical point of view, the crucial fact uniting these two strands is their reliance on particular technologies that function within – and only within – specifically designed spaces that are public (or publicly funded), rational and facilitating a mutually structuring relationship between individuals and society at large. Let us initially turn to the arguably more influential concept of the 'public sphere' before briefly turning to work within historical geography deriving from the second strand, which is centrally associated with the term 'governmentality'.

The practice of secularization mentioned earlier in this chapter, as well as related processes associated with the Enlightenment and its direct political consequences (of which the American and French Revolutions are perhaps the best-known manifestations), required the development of new spaces for the free expression of thoughts, ideas and actions. The formulation of this link between political theory, nascent republican and ultimately democratic practices and geographical sites owes a lot to Jürgen Habermas's *Structural Transformation of the Public Sphere*, originally published in 1962 and in a reliable English translation in 1991. Habermas's book and the ideas expressed therein have influenced historically orientated research like few other books in its generation. A historically cognizant sociologist by training and indebted himself to the writings of the Frankfurt School of Social Theory, Habermas successfully linked the emergence of new social formations in the eighteenth century, the age of Enlightenment – which would prove to be key to the development of capitalism in the nineteenth and twentieth centuries – to everyday spaces like cafés, (literary) salons or Masonic lodges. It was in these spaces that the bourgeoisie was first able effectively to communicate – and thereby put into practice – interests of their own design and making, eventually leading to the creation of information networks operating at a host of scales (Ogborn, 2007). Instrumental for the realization of such business interests was the parallel development of political freedoms and influence, setting in motion an incremental – and quintessentially modern – struggle over political rights that continues to the present day.

The appeal of Habermas's work to historical geographers is obvious and resides in the bringing together of a normative reading of modern history with a highly spatialized analysis of concrete historical practices (Howell, 1993). While broadly welcomed by geographers across a range of specializations, the problem with the Enlightenment spaces mentioned was, of course, their exclusive nature: originally attaching to the emerging 'public sphere' of initially Western (actually North-Western) Europe, access to these spaces was structurally restricted by class, gender and race (Bondi and Domosh, 1998; Cooper et al., 2000); geography, too, in the form of a thoroughly urban bias built into notions of the public sphere, limited access to 'public spheres' (Goheen, 1998; Kincaid, 2003) and here in particular to 'the street' (Goheen, 1993). Just as important has proven to be a somewhat limiting conceptualization of the link between the 'public' nature of progressive politics in Habermas and its genesis in necessarily private concerns and desires. Not only did these latter prove to be anything but a clear-cut expression of independently minded and developing interests – thus implying, rather than founding, the modern political individual as the central actor of modern politics. In addition, the notion of 'the private' was increasingly seen to be linked to 'the public' only where and if it gave birth to economically pertinent activities (McGurty, 1998; but see Baydar, 2003 for an interesting reading of key texts pertinent to the 'public'–'private' differentiation). Other forms of socially relevant actions and practices – from child rearing to education and social networking – did not initially seem of importance in this analytical framework, thereby creating a highly particular notion of 'the political'.

Not surprisingly, the problems that had characterized an initially tempting match between the public sphere and its inherent – if historically contingent – progressive nature thus continue to be the focus of work in historical geography. Planning, for instance, and the popularization of its achievements, structured the interface between public and private realms, which Habermas defined as key in the formation of modern democratic societies, in particularly gendered and thus non-universal manners (Lloyd and Johnson, 2004; for a related discussion of the gendering of ostensibly universal design, see Leslie and Reimer, 2003, as well as Jerram, 2006). This lack of universality in the realization of planned spaces often brought forth a rather complex relationship between 'plan', 'planner' and 'planned' that contradicts any simplistic reading of modernity as a process (or indeed 'project', see Habermas, 1987: xix) leading to participatory forms of governance (Llewellyn, 2004).

Modernity and Governmentality

In addition to the Habermasian lines of thought and research discussed above, historical geographers interested in modernity and 'modern' politics have also drawn inspiration from the work of Michel Foucault (Philo, 1992, 2000; Hannah, 1993). In truth, Foucault's entire oeuvre offers plenty of ideas and historically resonant material for geographers to engage with but, more than most of his concepts, it is the term 'governmentality' that has inspired work in geography, broadly construed. 'Governmentality' re-inscribes the site of 'the political' by bridging the distance between individual and society through analyses (historical and otherwise) of such socially relevant practices that contribute to the realization of government policies. Central amongst these are instruments and techniques that locate the bodies of individual citizens in spaces deemed appropriate by the state, be it by rendering nomadic practices illegitimate (Hannah, 1993), by translating urban chaos into legible spaces through the numbering of houses (Rose-Redwood, 2006; Strohmayer, 2009), the disciplining of bodies (Crowley and Kitchin, 2008; Hannah, 2009) and related practices both in Western societies and colonial settings (Legg, 2006). Due to its explicit involvement in the positioning (or 'geo-coding', see Rose-Redwood, 2006) and conduct of historical bodies in space, the concept of governmentality is also often associated with the notion of 'biopolitics' in studies of historical continuity and change (Legg, 2005; Gandy, 2006; Nally, 2008; see also Chapter 20).

The key both to Foucault's own work on governmentality and the work of historical geographers following his lead is the documentation and analysis of modernity as a process marked by an often colossal expansion of state-run techniques aimed at optimizing the act of governing in the form of new bureaucracies, novel institutions and original mentalities to match (and emanating from) same. The hinge linking individuals and society in many historical studies is the rise and acceptance of the role of experts in modern societies (Hannah, 2000; see also Braun, 2000 for an interesting case study). Taken together, the techniques employed all lend structure to the modernizing of societies, routinize the act of governing and ultimately create a sense of complicity between individuals and state.

Conclusion

To summarize, readers would not be wrong to recognize two sides to the coin of modernity in the two main 'political' strands of inquiry detailed above: one focusing the realization of individual freedoms, the other on the historical formation of social cohesions. However, both analyses centred around the 'public sphere' and 'governmentality' explicitly hone in on key conditions of possibility of social progress and should thus be seen as complementing, rather than contradicting one another.

We ought to note in closing that the centrality accorded to 'modernity' and associated themes within geography in recent years has finally had the unfortunate – and largely unintended – consequence of a 'considerable narrowing of the time periods that inform [human geography's] empirical and conceptual studies' (Jones 2004: 288). Historical geographers may be more immune to such claims but even here, with the notable contemporary exception of Keith Lilley, theoretically informed works on say medieval landscapes appear to have largely vanished from the pages of geographical journals. Thankfully, this reduction in the scope of temporal curiosity during the last decade or two has been matched by a welcome expansion in the scope of geographical curiosity beyond predominantly English-speaking spaces and epochs. Hence the sheer breadth and diversity of the work that informs the many chapters of the present book, all of which are testimony of a vibrant research culture characterizing human geography in the twenty-first century.

KEY POINTS

- Modernity is an oft-invoked but imprecise term in historical geography.
- The term is used mostly in conjuncture with other concepts, all aimed at interpreting and analysing processes of social, cultural and economic change.
- In addition, 'modernity' carries the burden of explaining the rise of progressive and increasingly inclusive forms of governance.
- Key in this latter attempt has proven to be the work of Jürgen Habermas, whose *Structural Transformation of the Public Sphere* is the seminal text for many historical attempts to trace the genesis of modern politics.
- In addition to the 'public'–'private' difference, historical geographers have also turned to the notion of 'governmentality' to analyse key political aspects associated with modernity.

References

Bassett, K. (2008) 'Thinking the event: Badiou's philosophy of the event and the example of the Paris Commune', *Environment and Planning D: Society and Space*, 26: 895–910.

Baydar, G. (2003) 'Spectral returns of domesticity', *Environment and Planning D: Society and Space*, 21(1): 27–45.

Bondi, L. and Domosh, M. (1998) 'On the contours of public space: a tale of three women', *Antipode*, 30(3): 270–289.

Bonnemaison, S. (1998) 'Moses/Marianne parts the Red Sea: allegories of liberty in the bicentennial of the French Revolution', *Environment and Planning D: Society and Space*, 16(3): 347–365.

Braun, B. (2000) 'Producing vertical territory: geology and governmentality in late Victorian Canada', *Cultural Geographies*, 7(1): 7–46.

Clayton, D. (2000) 'The creation of imperial space in the Pacific Northwest', *Journal of Historical Geography*, 26(3): 327–350.

Collier, P. and Inkpen, R. (2003) 'The Royal Geographical Society and the development of surveying 1870–1914', *Journal of Historical Geography*, 29(1): 93–108.

Cooper, A., Law, A., Malthus, J. and Wood, P. (2000) 'Rooms of their own: public toilets and gendered citizens in a New Zealand city, 1860–1940', *Gender, Place and Culture*, 7(4): 417–433.

Crowley, U. and Kitchin, R. (2008) 'Producing "decent girls": governmentality and the moral geographies of sexual conduct in Ireland (1922–1937)', *Gender, Place and Culture*, 15(4): 355–372.

Gandy, M. (2006) 'Zones of indistinction: bio-political contestations in the urban arena', *Cultural Geographies*, 13(4): 497–516.

Goheen, P. (1993) 'The ritual of the streets in mid-19th-century Toronto', *Environment and Planning D: Society and Space*, 11(2): 127–145.

Goheen, P. (1998) 'Public space and the geography of the modern city', *Progress in Human Geography*, 22(4): 479–496.

Graham, B. and Shirlow, P. (2002) 'The Battle of the Somme in Ulster memory and identity', *Political Geography*, 21(7): 881–904.

Habermas, J. (1987) *The Philosophical Discourse of Modernity: Twelve Lectures*. Cambridge, MA: MIT Press.

Habermas, J. (1991) *The Structural Transformation of the Public Sphere*. Cambridge, MA: MIT Press.

Hannah, M. (1993) 'Space and social control in the administration of the Oglala Lakota ("Sioux"), 1871–1879', *Journal of Historical Geography*, 19(4): 412–432.

Hannah, M. (2000) *Governmentality and the Mastery of Territory in Nineteenth-century America*. Cambridge: Cambridge University Press.

Hannah, M. (2009) 'Calculable territory and the West German census boycott movements of the 1980s', *Political Geography*, 28(1): 66–75.

Harvey, D. (1979) 'Monument and myth: the building of the Basilica of the Sacred Heart', *Annals of the Association of American Geographers*, 69(3): 362–381.

Heffernan, M. (2001) 'History, geography, and the French national space: the question of Alsace-Lorraine', *Space and Polity*, 5(1): 27–48.

Helleiner, E. (1999) 'Historicizing territorial currencies: monetary space and the nation-state in North America', *Political Geography*, 18(3): 309–339.

Howell, P. (1993) 'Public space and public sphere: political theory and the historical geography of modernity', *Environment and Planning D: Society and Space*, 11(3): 303–322.

Jerram, L. (2006) 'Kitchen sink dramas: women, modernity and space in Weimar Germany', *Cultural Geographies*, 13(4): 538–556.

Jones, R. (2004) 'What time human geography?', *Progress in Human Geography*, 28(3): 287–304.

Kincaid, S. (2003) 'Democratic ideals and the urban experience', *Philosophy and Geography*, 6(2): 145–152.

Konvitz, J. (1990) 'The nation-state, Paris and cartography in eighteenth- and nineteenth century France', *Journal of Historical Geography*, 16(1): 3–16.

Legg, S. (2005) 'Foucault's population geographies: classification, biopolitics and governmental spaces', *Population, Space and Place*, 11(3): 137–156.

Legg, S. (2006) 'Governmentality, congestion and calculation in colonial Delhi', *Social & Cultural Geography*, 7(5): 709–729.

Leslie, D. and Reimer, S. (2003) 'Gender, modern design, and home consumption', *Environment and Planning D: Society and Space*, 21: 293–316.

Lilley, K. (2004) 'Mapping cosmopolis: moral topographies of the medieval city', *Environment and Planning D: Society and Space*, 22(5): 681–698.

Llewellyn, M. (2004) '"Urban village" or "white house": envisioned spaces, experienced places, and everyday life at Kensal House, London in the 1930s', *Environment and Planning D: Society and Space*, 22: 229–249.

Lloyd, J. and Johnson, L. (2004) 'Dream stuff: the postwar home and the Australian housewife, 1940–60', *Environment and Planning D: Society and Space*, 22: 251–272.

Maddrell, A. (1998) 'Discourses of race and gender and the comparative method in geography school texts 1830–1918', *Environment and Planning D: Society and Space*, 16(1): 81–103.

McGurty, E. (1998) 'Trashy women: gender and the politics of garbage in Chicago, 1890–1917', *Journal of Historical Geography*, 26: 27–43.

Moran, D. (2006) 'Soviet cartography set in stone: the "map of industrialisation"', *Environment and Planning D: Society and Space*, 24(5): 671–689.

Nally, D. (2008) '"That coming storm": the Irish Poor Law, colonial biopolitics, and the Great Famine', *Annals of the Association of American Geographers*, 98(3): 714–741.

Ogborn, M. (1998) *Spaces of Modernity: London's geographies, 1680–1780*. New York: Guilford Press.

Ogborn, M. (2007) *Indian Ink: Script and Print in the Making of the English East India Company*. Chicago: University of Chicago Press.

Philo, C. (1992) 'Foucault's geography', *Environment and Planning D: Society and Space*, 10(2): 137–161.

Philo, C. (2000) '"The Birth of the Clinic": an unknown work of medical geography', *Area*, 32(1): 11–19.

Power, M. (2001) 'Geo-politics and the representation of Portugal's African colonial wars: examining the limits of "Vietnam Syndrome"', *Political Geography*, 20(4): 461–491.

Rose-Redwood, R. (2005) 'Governmentality, geography and the geo-coded world', *Progress in Human Geography*, 30(4): 469–486.

Rose-Redwood, R. (2006) 'Indexing the great ledger of the community: urban house numbering, city directories, and the production of spatial legibility', *Journal of Historical Geography*, 34(2): 286–310.

Sparke, M. (1998) 'A map that roared and an original atlas: Canada, cartography, and the narration of nation', *Annals of the Association of American Geographers*, 88(3): 463–495.

Strohmayer, U. (1995) 'From Weimar to Nuremberg: Social legitimacy as a spatial process in Germany, 1923–1938', in G. Benko and U. Strohmayer (eds) *Geography, History and Social Science*. Dordrecht: Kluwer Academic Publishers, pp. 143–170.

Strohmayer, U. (2009) 'Bridges: different conditions of mobile possibilities', in P. Merriman and C. Cresswell (eds) *Mobilities: Practices, Spaces, Subjects*. Farnham: Ashgate.

Tang, C. (2008) *The Geographic Imagination of Modernity: Geography, Literature, and Philosophy in German Romanticism*. Stanford, CA: Stanford University Press.

Toulmin, S. (1990) *Cosmopolis: The Hidden Agenda of Modernity*. New York: Macmillan.

FURTHER READING

Braun, B. (2000) 'Producing vertical territory: geology and governmentality in late Victorian Canada', *Cultural Geographies*, 7(1): 7–46.

Hannah, M. (1993) 'Space and social control in the administration of the Oglala Lakota ("Sioux"), 1871–1879', *Journal of Historical Geography*, 19(4): 412–432.

Harvey, D. (1979) 'Monument and myth: the building of the Basilica of the Sacred Heart', *Annals of the Association of American Geographers*, 69(3): 362–381.

Howell, P. (1993) 'Public space and public sphere: political theory and the historical geography of modernity', *Environment and Planning D: Society and Space*, 11(3): 303–322.

Nally, D. (2008) '"That coming storm": the Irish Poor Law, colonial biopolitics, and the Great Famine', *Annals of the Association of American Geographers*, 98(3): 714–741.

Section 7
Beyond the Border

19 GLOBALIZATION

David Nally

Introduction

'All major social forces have precursors, precedents and sources in the past', writes anthropologist Arjun Appadurai (1996: 2). This is indeed the case and globalization is no exception. Although the term itself is of recent origin – the verb to 'globalize' first appeared in the 1960s – most scholars recognize that the phenomenon has a far longer history. Difficulties arise, however, when efforts are made to delineate the historical geography of globalization more definitively. Does globalization originate with the first movement of peoples out of Africa? Or much later at the end of the thirteenth century when the Venetian merchant Marco Polo returned home with outlandish tales of Kublai Khan and the hidden marvels of Asia – tales which reawakened trade between Europe and the 'Far East' and fired the imagination of another young explorer, a Genovese native named Christopher Columbus, who in attempting to find an alternative route to the Indian subcontinent 'discovered' the Americas and kick-started a great wave of European colonial expansion (see also Chapters 1 and 2)? Perhaps the seeds of globalization were sown four centuries later, in the late nineteenth century, when a spirit of internationalism 'produced universal exhibitions, universal time ... plans for a universal language ... the International Geographical Union and proposals for an international map of the world' (Cosgrove, 2001: 207)?

However, framing the question in this way presupposes that there is a single, uniform 'ur-history' of globalization replete with a founding story and an organic, evolutionary trajectory. A contrasting approach is to conceptualize globalization as a *process* rather than an event. The latter strategy is adopted by historian A.G. Hopkins (2002: 3) who describes globalization as a 'series of overlapping and interacting sequences rather than a succession of neat stages'. This decision to work through themes in an interfoliated fashion, grasping not just the complementarity and interdependence of anchor concepts like 'modernity', 'globalization' and 'capitalism', but also paying heed to how ideas and practices continually resurface beyond the setting in which they were first elaborated, is remarkably similar to Edward Said's (2000: 186) call for a 'contrapuntal' reading of history. A de-centred, contrapuntal reading of globalization is one sensitive to the fact that many of the factors deemed unique to contemporary globalization – cross-border movements of peoples, the extension of communication networks, flows of goods and services, the spread of world religions, the deepening of market relations, the creation of international institutions, the ascent of cosmopolitan spirit – have existed for centuries, albeit with different degrees of intensity and affect and often with contradictory outcomes.

According to Andre Gunder Frank (1993), for example, the foundations for a global-scale world system began in the Bronze Age, with the emergence of city states in southern Mesopotamia (or modern-day Iraq where intensive agriculture, proto-industries and the growing complexity of regional trade led to the invention of the first written systems) and an irrigated Egyptian Empire, which at its zenith extended north to modern Syria and as far south as the Republic of Sudan (Gills and Thompson, 2006). In contrast, the eminent sociologist Immanuel Wallerstein (1974) believes that the contemporary global order reflects structural changes initiated in the early colonial period when a capitalist world economy first took root (see also Chapter 16). For Wallerstein, the emergence of a global 'world system' is an inherently spatial process in which 'core countries' were able to develop economically and politically through the extraction of surplus from 'peripheral countries' (Taylor and Flint, 2000). Rather than see these different theorizations of globalization as fundamentally irreconcilable, Hopkins (2002: 3) points out that globalization is a process that takes different forms, including 'archaic, proto, modern, and post-colonial' expressions.

Seen this way, Gunder Frank's work lays emphasis on what Hopkins would call 'archaic globalization', whereas the term 'proto-globalization' more accurately captures Wallerstein's stress on the establishment of Western European hegemony after the colonization of the 'New World'. For others, such as Kenneth Pomeranz, European colonial ventures in the New World (an 'ecological windfall' – or more accurately a ecological subsidy – that 'obviated the need to manage land intensively') together with the 'geographical accident' of extensive coal reserves in the Old World (occasioning a great 'energy revolution') combined to birth a truly global economy characterized above all by a 'great divergence', from 1750, between Europe and East Asia (Pomeranz, 2000: 13, 16, 23). In Enrique Dussel's unique genealogy, globalization is the outcome of new 'reflexive consciousness' that saw Europe reconstitute itself at the centre of a new 'planetary paradigm' (Dussel, 1998: 4) made possible by the 'discovery' of the New World. It was Amerigo Vespuccio, in 1503, who first suspected that the Spanish had actually found a new continent. Before this moment the fundamental structure of globalism was impossible to conceive. For this reason Dussell credits Amerigo Vespuccio with being 'existentially ... the first Modern' – the first truly global citizen. For Dussel (1998: 10), 'modern subjectivity develops spatially'.

Flattening the World

To focus too exclusively on the origin(s) of globalization can obscure other aspects of the process that warrant sustained scrutiny. In a series of publications Manfred B. Steger (2005, 2009a, 2009b) has called attention to the ideological dimensions of the globalism. Steger suggests that theories of globalization are often underwritten by strong normative claims. These include the belief that globalization is synonymous with economic growth, which is best facilitated by the liberalization of markets. Second, while globalization can speed up or slow down, progress is inexorable and unidirectional. Third, globalization benefits everyone. Finally, it is often assumed that globalization advances the spread of democracy and freedom. Seen this way, globalization interpellates us as subjects (Althusser, 1971), generating a system of assumptions that we subscribe to and desire (Boltanski and Chiapello, 2005). In other words, globalization becomes a societal aspiration.

Many of the ideological traits captured by Steger are evident in the journalist Thomas L. Friedman's (2005) award-winning account of globalization, *The World is Flat*. According to Friedman there have been three distinct but related historical periods of globalization, the first of which he calls Globalization 1.0. Globalization 1.0 began when Columbus set sail and by an accident of history, discovered the Americas and opened up trade between the 'Old World' and the 'New'. For Friedman this 'discovery' created the conditions for 'knitting the world together' and driving the first real wave of 'global integration'. The key engine of change in Globalization 1.0 was the productive harnessing of power – horsepower, wind power and later steam power – which Friedman views as central to facilitating what he terms the great 'Age of Discovery'. In this first great era of global integration Friedman suggests that the world shrank from a size large to a size medium. In other words, the movement of goods and people brought far-flung and very different regions of the world much closer together.

The second great era, Globalization 2.0, lasted between 1800 and 2000. In this era the world shrank from a size medium to a size small. According to Friedman, the key agents of change in Globalization 2.0 are 'multinational companies'. Spearheaded by the Industrial Revolution and the expansion of the Dutch and English joint-stock companies, multinationals scanned the now shrinking globe for new markets and 'untapped' reserves of labour, which could be 'captured' and brought into the fold of what Friedman sees as a uniquely European system of capitalist production. Whereas in Globalization 1.0, global integration was powered by the falling transportation costs (especially rail and steam), in Globalization 2.0 integration was powered by falling telecommunications costs. According to Friedman, this latest convergence is qualitatively different to any previous era of Globalization: in Globalization 1.0 European countries were the main actors; in Globalization 2.0, Dutch, English and eventually American corporations powered global integration; whereas today, in Globalization 3.0, it is 'empowered individuals' driving the global economy. So long as one has a computer – and an internet connection – it is possible to 'plug and play' – his words – in the global economy. Crucially, through all stages

Table 19.1 Flat-world ontology

	Period	Technology	Medium	Shrinkage
Globalization 1.0	1492–1800	Horsepower, wind power, steam power	States	Large world to a small world
Globalization 2.0	1800–2000	Telegraph, telephones, PCs, World Wide Web	Multinational Corporations	Small world to a tiny world
Globalization 3.0	2000–present	PCs, fibre-optic cable, workflow software	'Empowered Individuals'	Flat world

of globalization integration is driven by the advanced regions of the global North for the greater good of humanity. A flattening world is *ipso facto* a more equal world. To draw on an old adage, globalization is 'the rising tide that lifts all boats'.

Friedman's arguments – which we can characterize, following Sparke's critical intervention, as a 'flat-world ontology' (cf. Sparke, 2007a, 2007b) – can be critiqued at a number of levels. First, the argument is not just descriptive; it is *prescriptive*. As John Abraham (1991) has written (albeit in a different context) 'arguments about how we ought to develop ... are generally derived from a view of how we have developed to date'. In other words, Friedman is not only telling a story about past development; he is saying that globalization – powered by unbridled capitalism – is now, and has always been, an 'empowering' process, a great leveller of human inequalities, and thus it should be unequivocally supported in the present. The flat-world ontology is, however, deeply (and disturbingly) amnesiac about present and past inequalities. Although there is little doubt that global wealth, measured in terms of per capita income, has increased greatly during the period of capitalist development, it is also the case that those gains are concentrated in fewer and fewer hands. Indeed the 'gap between the centre and the periphery', as John Bellamy Foster (1999: 20) points out, 'grew from 1:1 in 1750 to 4:1 in 1930 to 7:1 in 1980'. In the 1990s the net worth of the 358 richest people in the world was equivalent to the cumulative income of the poorest 45 per cent of the world's population (around 2.3 billion people) and in 2006 the UN reported than the richest 2 per cent of adults control more than 50 per cent of global wealth (Davies, 2008; Harvey 2009: 58–60). For this reason David Harvey (2005) has suggestively argued that globalization is a means to reconstitute and entrench class power on the world stage.

A second limitation of the 'flat world' thesis is its *teleological* nature – it assumes that history has an intrinsic purpose: the West has been, and will be, the primary engine for developing the poorer nations of the global South. As an acolyte of Adam Smith, Friedman believes that global 'growth' is synonymous with global 'development', and, moreover, that human innovation, especially Western innovation, drives large-scale social, political and economic transformations. Geographer J.M Blaut (1993) memorably described this blinkered style of thinking as 'geographical diffusionism'. Diffusionists such as Friedman fathom a world with a 'permanent geographical centre and permanent periphery: an Inside and an Outside. Inside leads, Outside lags, Inside innovates,

Outside imitates' (Blaut, 1993: 1). According to this stylized 'Eurocentric' narrative of globalization, 'It all happened in Europe: first came the Renaissance, the Enlightenment and the Industrial Revolution, and this led to a massive rise in living standards in the West. And now those great achievements of the West are spreading to the world' (Sen, 2006: 125). Not only does 'geographical diffusionism' disregard the accomplishments of non-Western peoples (or consign those achievements to a mere footnote in the larger annals of Euro-American progress); it also cultivates a sense of 'moral narcissism' (Ignatieff, 1999: 6), which if unchecked, can lead to a politics of aversion, bigotry and violence (Amin, 2012).

Globalization and Neoliberal Hegemony

In some accounts 'globalization' is simply a means of presenting 'Western values' as a universal good and common aspiration. For example, in *The End of History and the Last Man* (1992) the American neoconservative philosopher Francis Fukuyama argued that the fall of the Berlin Wall and the collapse of the Soviet Union marked the end of any viable alternative to free-market capitalism. In Fukuyama's view, history had reached a natural terminus and 'the future would now be simply more of the present' (Eagleton, 2011: 6). Fukuyama's book powerfully captured the mood of free-market triumphalism that dominated discussions of global change in the 1980s and played a formative role in defining US global ambition in the following decades. Although the principle of free markets has long been considered a defining feature of American foreign policy (Smith, 2005; Westad, 2007), the sudden demise of the Soviet Union allowed the US to position its own strategic priorities as universal values – in this way 'national security' was mapped outwards onto a contested global sphere (Dodds, 2007). The naturalism of the 'first world' was asserted via the material and discursive subjugation of the 'second' (Goldberg, 2003).

Pushing American borders 'out into the world,' as Neil Smith (2005: 202) describes it, is accomplished in myriad ways. The institutions governing the global economy – the World Trade Organization, the International Monetary Fund and the World Bank – press a narrow set of economic policies (free markets, privatization, monetarism, financial de-regulation, export-focused growth, the privatization of public assets and the rolling back of the welfare state) that bear the imprint of US national interests and are enforced globally via legal arbitrage and debt servitude (George, 1992; Hoogvelt, 2001; Harvey, 2005; Peet, 2009). Meanwhile Hollywood, corporate advertising, MTV and web-based media establish America as a 'cultural hegemon' – a 'superculture' as well as a 'superpower' (Meinig, 2004) – a nation whose beliefs and values unquestionably define the global *zeitgeist* (see also Chapter 15).

Raw military power is of course another means of pursuing global ambition. For Neil Smith (2005: ix) the ongoing global 'war on terrorism' is simply the 'continuation of globalisation by military means' (Chomsky, 2003; Kiernan, 2005; Hobsbawm, 2007; Kearns, 2009). Presented as struggle for freedom over repression, the US-led offensive established American control over some of the most productive and lucrative oilfields in the Middle East (Retort, 2005) and paved the way for the reconstruction of Iraqi society (Chandrasekaran, 2006). By the time power was handed over to the interim government, the US had undertaken the liberalization of tax, trade and investment laws to allow for the piecemeal sale of Iraq's assets to the highest corporate bidder. This was no small accomplishment since the Iraqi constitution expressly prohibited the privatization of vital economic assets and the Geneva Conventions, established in the aftermath of World War II, prevent occupying forces from interfering with the legal provisions in an occupied state. Brushing aside these restrictions a legal instruction, known as Order 39, codified the privatization of state-owned resources, legalized the foreign ownership of Iraqi banks, mines and factories, removed barriers for foreign direct investment and permitted the repatriation of profits out of Iraq (Klein, 2007: 345; Harvey, 2009: 55–56). Author and critic Naomi Klein (2007: 328, 351) describes how Iraq was 'transformed into a cutthroat capitalist laboratory', a place where combating terrorism and extending capitalism dovetailed in a unified political project. The term 'Washington Consensus' is often used to describe those within the US administration who believed that the government of Iraq provided a template for global domination (see also Chapter 6).

Goodbye State?

The current era of globalism has not led to the erosion of the state (*pace* Friedman), nor has it led to the portentous future-without-an-alternative announced by Fukuyama. Capitalist activity, as Harvey (2003) reminds us, has always relied on the institutions of the state to foster the enabling conditions for the circulation of people, capital and goods. At the turn of the seventeenth century, for instance, the English government formed the East India Trading Company, granting it special exemptions and trading monopolies to wrestle control of markets in tea, cotton, silk and opium. In 1602 the Dutch responded by forming the Dutch East India Company (VOC), using trade restrictions and state monopolies to control commodity markets in South Asia (Ogborn, 2000, 2008). It is true today that a more 'complex, mutating and stratified' (Duffield, 2001: 12) assemblage of non-governmental organizations, international financial institutions, United Nations agencies and financial corporations comprise the field of global governance, but these 'emerging political complexes', as Mark Duffield terms them, still require the harmonizing powers of states to both operate in and manage a liberal market environment. The development of capitalism, seen as global historical and geographical process (Arrighi, 1994), must permanently deploy a 'visible hand' to generate unceasing profit and accumulation (Harvey, 2009: 99–100).

But this is only part of the story for, as Karl Polanyi (2001: 76) famously noted, the move to commodify everything (in other words, 'globalize' capitalist relations) *inevitably* engenders resistance:

> To allow the market mechanism to be the sole director of the fate of human beings and their natural environment ... would result in the demolition of society. For the alleged commodity "labor power" cannot be shoved about, used indiscriminately, or even left unused, without affecting also the human individual who happens to be the bearer of this particular commodity. In disposing of man's labour power the system would, incidentally, dispose of the physical, psychological, and moral entity "man" attached to that tag. Robbed of the protective covering of cultural institutions, human beings would perish from the effects of social exposure; they would die as victims of acute social dislocation through vice, perversion, crime and starvation. Nature would be reduced to its elements, neighborhoods and landscapes defiled, rivers polluted, military safety jeopardized, the power to produce food and raw materials destroyed.

In short, to permit the market to be the 'sole director' of material world would be to accept the obliteration of life itself. For this reason Polanyi (2001) theorized that the attempt to globalize capitalist relations would always meet and clash with counter-movements designed to temper market mechanisms, 're-embedding' the latter in the lived world of human–environmental relations. In the nineteenth century, during the period that Hopkins would term 'modern globalization', efforts to extend market relations were met with strong social movements that agitated for more progressive conditions for labourers, including restrictions on the number of working hours (cf. Kearns, 2009).

Anti-globalization and its Discontents

These early populist movements – sometimes collectively referred to as 'globalization from below' – are important antecedents to the grassroots movements that challenge the power of global capitalism today. These 'experiments against contemporary sources of inequality', to use Ash Amin and Nigel Thrift's (2013) term, include everything from food sovereignty, open sourcing and consumer boycotts to financial transaction taxes, cooperative enterprises, indigenous development initiatives and sustainable livelihoods. Such heterogeneous practices are often swept together and labelled the 'anti-globalization' movement, but as Amartya Sen (2006: 124) rightly points out, the 'anti-globalization critique is perhaps the most globalized moral movement in the world today' – proof, if ever it were needed, that Fukuyama's notion that the present is a condition for which there is no alternative is ideologically motivated and unrepresentative of the counter-currents that militate against the forces of privilege and injustice operating around the world (Hoogvelt, 2001).

Sen also adds, however, that so-called anti-globalization protestors often unwittingly, and certainly unhelpfully, *equate* injustice with globalization itself. 'It is a mistake', he says, 'to see deprivations and divided lives as penalties of globalisation, rather than as failings of social, political, and economic arrangements, which are entirely contingent and not inescapable companions of global closeness' (Sen, 2006: 122). Anti-globalists also mistakenly assume 'that there is such thing as "*the* market outcome", no matter what rules of private operation, public initiatives, and nonmarket institutions are combined with the existence of markets' (Sen, 2006: 136). Seeing capitalist markets as made – a point made long ago by Polanyi (2001) and recently reiterated in two luminous studies (Chang, 2008; Harcourt, 2011) – opens up the possibility that they might be *made differently* (George, 2004; Barnes, 2006). 'The market economy does not work alone in *globalized* relations,' concludes Sen, 'indeed it cannot operate alone even *within* a country.' It follows that

> the nature of market outcomes are massively influenced by public policies in education and literacy, epidemiology, land reform, micro-credit facilities, appropriate legal protections etc., and in each of these fields there are things to be done through public action that can radically alter the outcome of local and global economic relations. (Sen, 2006: 138)

Conclusion

Is there a role for historical geography in fostering the kind of critical public action that Amartya Sen describes? Indeed there is, but that role must involve moving beyond a narrow focus on the origins of globalization (important though that subject is) to consider the normative claims that underpin mainstream accounts of globalism, whether that be Friedman's (2005) Panglossian vision of planetary integration, Fukuyama's (1992) platitudinous statements about 'the end of history' and the birth of universal and homogeneous 'Last Man,' or the kind of fire-brand narratives supplied by Samuel Huntington (1996) and Robert Kaplan (2000) suggesting that civilizational confrontation will inevitably ensue as irreconcilable cultures are forced to share a narrowing global stage. Each of these theorizations is a violence visited on diversity and plurality, even though they sometimes masquerade as celebrations of both. A critical historical geography is well placed to point out such shortcomings.

KEY POINTS

- There is no agreed consensus on the origins of a global-scale world system, but all scholars agree that contemporary globalization has important sources and precedents in the past.
- Historical geographers now pay greater attention to the ideological dimensions of globalization, including the establishment of hegemonic status through colonial and later imperial relations.
- The development of capitalism, seen as a global historical and geographical process, has profoundly shaped contemporary globalization.
- Historical geographers have challenged the view that globalization is a natural force and a universal good. History shows that globalization has produced 'winners' and 'losers'. Globalization is unlikely to mark the 'end of history' – or indeed the 'end of geography'.

References

Abraham, J. (1991) *Food and Development: The Political Economy of Hunger and the Modern Diet*. London: Kogan Page.

Althusser, L. (1971) *Lenin and Philosophy and Other Essays* (trans. B. Brewster). London: New Left Books.

Amin, A. (2012) *Land of Strangers*. Cambridge: Polity Press.

Amin, A. and Thrift, N. (eds) (2013) *Arts of the Political*. Durham, NC: Duke University Press.

Appadurai, A. (1996) *Modernity at Large: Cultural Dimensions of Globalization*. Minneapolis: University of Minnesota Press.

Arrighi, G. (1994) *The Long Twentieth Century: Money, Power, and the Origins of Our Times*. London: Verso.

Barnes, P. (2006) *Capitalism 3.0: A Guide to Reclaiming the Commons*. San Francisco: Berrett-Koehler.

Blaut, J.M. (1993) *The Colonizer's Model of the World: Geographical Diffusionism and Eurocentric History*. New York: Guilford Press.

Boltanski, L. and Chiapello, E. (2005) *The New Spirit of Capitalism* (trans. G. Elliot). London: Verso.

Chandrasekaran, R. (2006) *Imperial Life in the Emerald City: Inside Iraq's Green Zone*. New York: Alfred A. Knopf.

Chang, H.J. (2008) *Bad Samaritans: The Myth of Free Trade and the Secret History of Capitalism*. New York: Bloomsbury.

Chomsky, N. (2003) *Hegemony or Survival: America's Quest for Global Dominance*. New York: Hamish Hamilton.

Cosgrove, D. (2001) *Apollo's Eye: A Cartographic Genealogy of the Earth in the Western Imagination*. Baltimore, MD: Johns Hopkins University Press.

Davies, J.B. (ed.) (2008) *Personal Wealth from a Global Perspective*. Oxford: Oxford University Press.

Dodds, K. (2007) *Geopolitics: A Very Short Introduction*. Oxford: Oxford University Press.

Duffield, M. (2001) *Global Governance and the New Wars: the Merging of Development and Security*. London: Zed Books.

Dussel, E. (1998) 'Beyond Eurocentrism: the world-system and the limits of modernity', in F. Jameson and M. Miyoshi (eds) *The Cultures of Globalization*. Durham, NC: Duke University Press, pp. 3–31.

Eagleton, T. (2011) *Why Marx Was Right*. New Haven, CT: Yale University Press.

Foster, J.B. (1999) *The Vulnerable Planet: A Short Economic History of the Environment*. New York: The Monthly Review Press.

Frank, A.G. and Gills, B.K. (eds) (1993) *The World System: Five Hundred Years or Five Thousand?* (foreword by William H. McNeill). New York: Routledge.

Friedman, T.L. (2005) *The World is Flat: The Globalized World in the Twenty-First Century*. New York: Penguin.

Fukuyama, F. (1992) *The End of History and the Last Man*. New York: The Free Press.

George, S. (1992) *The Debt Boomerang: How Third World Debt Harms Us All*. London: Pluto Press.

George, S. (2004) *Another World Is Possible If ...* London: Verso.
Gills, B.K. and Thompson, W.R. (2006) *Globalization and Global History*. New York: Routledge.
Goldberg, D. T. (2003) *Racist Culture: Philosophy and the Politics of Meaning*. Cambridge, MA: Blackwell.
Harcourt, B.E. (2011) *The Illusion of the Free Markets: Punishment and the Myth of Natural Order*. Cambridge, MA: Harvard University Press.
Harvey, D. (2005) *A Brief History of Neoliberalism*. Oxford: Oxford University Press.
Harvey, D. (2009) *Cosmopolitanism and the Geographies of Freedom*. New York: Columbia University Press.
Hobsbawm, E.J. (2007) *Globalisation, Democracy and Terrorism*. London: Abacus.
Hoogvelt, A. (2001) *Globalization and the Postcolonial World: The New Political Economy of Development* (2nd edn). Basingstoke: Palgrave.
Hopkins, A. G. (ed.) (2002) *Globalization in World History*. New York: W. W. Norton.
Huntington, S. (1996) *The Clash of Civilizations and the Remaking of World Order*. New York: Simon & Schuster.
Ignatieff, M. (1999) 'The stories we tell: television and humanitarian aid,' *The Social Contract*, 10(1): 1–8.
Kaplan, R. (2000) *The Coming Anarchy: Shattering the Dreams of the Post Cold War*. New York: Vintage Books.
Kearns, G. (2009) *Geopolitics and Empire: The Legacy of Halford MacKinder*. Oxford: Oxford University Press.
Kiernan, V.G. (2005) *America: The New Imperialism*. London: Verso.

Klein, N. (2007) *The Shock Doctrine: The Rise of Disaster Capitalism*. London: Allen Lane.
Meinig, D.W. (2004) *The Shaping of America: A Geographical Perspective on 500 Years of History. Volume 4: Global America 1915–2000*. New Haven, CT: Yale University Press.
Ogborn, M. (2000) 'Historical geographies of globalisation', in B. Graham and C. Nash (eds) *Modern Historical Geographies*. London: Pearson, pp. 43–69.
Ogborn, M. (2008) *Global Lives: Britain and the World 1550–1800*. Cambridge: Cambridge University Press.
Peet, R. (2009) *Unholy Trinity: The IMF, World Bank and WTO* (2nd edn). London: Zed Press.
Polanyi, K. (2001) *The Great Transformation: The Political and Economic Origins of Our Times*. Boston: Beacon Press.
Pomeranz, K. (2000) *The Great Divergence: China, Europe, and the Making of the Modern World Economy*. Princeton NJ: Princeton University Press.
RETORT (2005) *Afflicted Powers: Capital and Spectacle in a New Age of War*. London: Verso.
Said, E. (2000) *Reflections on Exile and Other Essays*. Cambridge, MA: Harvard University Press.
Sen, A. (2006) *Identity and Violence: The Illusion of Destiny*. New York: W.W. Norton & Company.
Smith, N. (2005) *The Endgame of Globalization*. New York: Routledge.
Sparke, M. (2007a) 'Everywhere but always somewhere: critical geographies of the Global South', *The Global South*, 1(1/2): 117–126.

Sparke M. (2007b) 'Acknowledging responsibility for space', *Progress in Human Geography*, 31(3): 395–403
Steger, M.B. (2005) *Globalism: Market Ideology Meets Terrorism* (2nd edn). Lanham, MD; Oxford: Rowman & Littlefield.
Steger, M.B. (2009a) *Globalisms: The Great Ideological Struggle of the 21st Century* (3rd edn). Lanham, MD; Oxford: Rowman & Littlefield.
Steger, M.B. (2009b) *Globalization: A Very Short Introduction* (2nd edn). Oxford: Oxford University Press.
Taylor, P.J. and Flint, C. (2000) *Political Geography: World-Economy, Nation-State and Locality* (4th edn). New York: Prentice Hall.
Wallerstein, I. (1974) *The Modern World System: Capitalist Agriculture and the Origins of the European World-Economy in the Sixteenth Century*. New York: Academic Press.
Westad, O.A. (2007) *The Global Cold War: Third World Interventions and the Making of Our Times*. Cambridge: Cambridge University Press.

FURTHER READING

Appadurai, A. (1996) *Modernity at Large: Cultural Dimensions of Globalization*. Minneapolis: University of Minnesota Press.
Meinig, D.W. (2004) *The Shaping of America: A Geographical Perspective on 500 Years of History. Volume 4: Global America 1915–2000*. New Haven, CT: Yale University Press.
Ogborn, M. (2008) *Global Lives: Britain and the World 1550–1800*. Cambridge: Cambridge University Press.
Smith, N. (2005) *The Endgame of Globalization*. New York: Routledge.

20 GOVERNMENTALITY

David Nally

Introduction

In 1976 Michel Foucault (1926–1984), the renowned French philosopher and social theorist, was interviewed by a group of Marxist geographers associated with the radical French journal *Hérodote*. After first playing cagey with his interlocutors, Foucault admitted that geography was indeed 'the condition of possibility' for his enquiries. He went on to explain:

> The longer I continue the more it seems to me that the formation of discourses and the genealogy of knowledge need to be analyzed, not in terms of types of consciousness, modes of perception and forms of ideology, but in terms of tactics and strategies of power. Tactics and strategies deployed through implantations, distributions, demarcations, control of territories and organizations of domains which could well make up a sort of geopolitics where my preoccupations would link up with your methods ... Geography must indeed necessarily lie at the heart of my concerns. (Foucault, 2007a: 182)

Viewed retrospectively this interview offers a thumbnail sketch of the research questions that animate Foucault's later writings, particularly the redoubled focus on the 'tactics and strategies of power', but also an emerging sense that 'space is', as he subsequently put it, 'fundamental in any exercise of power' (Foucault, 1984: 252). These ideas crystallized in Foucault's celebrated study *Discipline and Punish: The Birth of the Prison* (published in French and translated into English around this time), a book that arguably marks a foundational moment in Foucault's thinking on the historical geography of liberal government. It is worth pausing, then, to consider Foucault's progression toward a distinctly spatial analytics of power and how this intellectual journey frames his subsequent reflections on the 'art of government' – that is, the exercise of control over individuals and whole populations, or what Foucault (1994a) succinctly termed 'governmentality'.

Disciplinary Power

Discipline and Punish begins with a gruesome description of the torture and public execution in 1757 of Robert-François Damiens for attempted regicide. The physical and public nature of Damiens' punishment is contrasted with the abstract and clinical practice of incarceration reserved for prisoners in more recent times. Foucault considered this profound shift in the exercise of punishment – from pain visited on the flesh to subtler forms of social control based on the ordering and directing of life – to be the defining characteristic of *disciplinary power*. Foucault's notion of disciplinary power owes much to the importance he placed on the Panopticon principle first articulated by the English philosopher and social reformer Jeremy Bentham (1748–1832). Bentham proposed the construction of a building that was specifically designed to enable an observer to inspect the inmates without the latter knowing whether or not they were being monitored. While Bentham had hoped to build a prison system based on the Panopticon (in fact those plans were never fully realized), he was also adamant that the principles underlying his building could be transposed to schools, hospitals, factories, workhouses, army barracks and asylums. In short, they could help configure the institutional structure of the liberal democratic state.

In Bentham's design Foucault recognized an attempt to redefine both the techniques and objectives of social government. The novelty of Bentham's proposal lay less in his principle of hierarchical observation than in his efforts to mould the environment, in this case the physical space of the prison, into 'an instrument and vector of power' (Foucault, 1979: 30) capable of operating on the body of the prisoner. The goal of 'normalizing' behaviour – that is, effecting a desired shift in the offender's general conduct and mode of action – would be achieved, not by administering recurrent cruelties, but by the careful supervision of the prisoner's milieu. Inside Bentham's Panopticon inmates would learn to discipline themselves; or, to phrase this differently, normative conduct could be *enforced without force*. 'All by a simple idea in architecture', as Bentham (1843: 66) famously put it.

This attempt to transpose the violence of correction into an art of government was viewed by Foucault to be a crucial feature of liberal rule. Instead of associating liberalism with the decision to deploy 'less cruelty, less pain, more kindness, more respect, more humanity', Foucault (1979: 16, 24) emphasized the degree to which the prisoner's body emerges as strategic site – an 'analysable' and 'manipulable' force; in short, a 'political instrument' – for realizing behavioural change. This political investment in the body (what Foucault would later call 'anatomo-politics') enabled power to penetrate 'into even the smallest details of everyday life' (Foucault, 1979: 198). Disciplinary apparatuses, as Foucault (1979: 138) went on to say, define 'how one may have hold over other's bodies, not only so that they may do what one wishes, but so that they may operate as one wishes, with the techniques, the speeds and the efficiency one determines'. Indeed, far from marking a diminution in the forces of subjugation, liberal government represented a new modality of power that rendered bodies both 'useful' and 'docile'. 'The "Enlightenment", which discovered the liberties', commented Foucault (1979: 222), 'also invented the disciplines.'

It is worth pausing here momentarily to stress some of the more striking features of Foucault's thesis on the birth of the modern prison. First, Foucault (1979: 222) is arguing that there is a 'dark side' to the modern penal code – that for all the piety and humanism that characterized public debate in the Age of Reform there remains, as it were, 'a trace of "torture" in the modern mechanisms of criminal justice' (Foucault, 1979: 17). Foucault's genealogy, then, is very much part of what Domenico Losurdo (2011) has recently termed a 'counter-history' of liberalism. Second, Foucault is clear that the proliferation of disciplinary power is a vital though overlooked element in the subordination of the worker under capitalism. Older methods of governing through domination, as Nikolas Rose (1999: 4) points out, meant 'crushing' rather than harnessing a person's capacity. Those methods of rule were less appealing to capitalists who now wished to rouse workers to create more and more surplus value through commodity production. Foucault continually stressed the point that disciplinary power was not 'repressive' – it stimulated and incited actions that were deemed desirable. For this same reason Foucault frequently described disciplinary systems as 'technologies': in much the same way that mechanical developments fostered new and more productive divisions of labour under industrial capitalism (MacKenzie, 1984), 'disciplinary technologies' generated new and more efficient ways of acting on populations. Elaborating on this very point, Foucault wrote: 'We frequently speak of the technical inventions of the seventeenth century – chemical, metallurgical technology – yet we do not mention the technical invention of this new form of governing man, controlling his multiplicity, utilizing him to the maximum, and improving the useful products of his labour ...' (2007b: 146). Here the political anatomy of the body is explicitly bound up with the political economy of capitalism; or as Foucault put it in *Discipline and Punish*, docility and utility were 'elements of the same system', designed to accomplish the triptych of outcomes, 'mildness-production-profit' (1979: 218–219). Seen this way 'disciplinary technologies' emerge as forces mediating 'our experience, our aspirations, our relations with ourselves and with others' (Rose, 1999: 11) – in short, as a force regulating the practice of everyday life.

The Art of Government

Philip Howell (2007: 300) points out that geographers have been 'quick to identify in Foucault's disciplinary genealogies a distinctive place for space'. Howell rightly identifies Felix Driver's (1993) analysis of the English Poor Law, and Chris Philo's (2004) study of the spatiality of madness in England and Wales, as important elaborations of Foucault's early work (cf. Mayhew, 2009). While Philo's study was more concerned with the link between 'power-knowledge' to be found in Foucault – how modes of calculation accrue a status of 'truth' and infuse political strategy – Driver was interested in mapping the regional variations in the disciplinary management of pauperism. Both works suggested the immense richness of Foucault's ideas and how they might be put to use in historical research. The increased focus on Foucault's political and philosophical writing also led to an engagement with the work that followed from *Discipline and Punish*. Most notably in *The History of Sexuality* – especially the final section of that study – Foucault signalled his interest in defining more carefully the problematic of government, and in particular what he termed the 'governmentalization of the state' (1980: 221). This refinement was driven by his sense that his earlier studies focused too narrowly on how power acted on the individual and, consequently, he had left to one side the issue of how populations are targeted, managed and regulated as *collective* phenomena. In his well-known essay 'Governmentality,' for example, Foucault sought to explain how, from at least the eighteenth century, government took on the task of administering life itself:

> The things ... with which government is to be concerned are in fact men, but men in their relations, their links, their imbrication with those things that are wealth, resources, means of subsistence, the territory with its specific qualities, climate, irrigation, fertility, and so on; men in their relation to those

> other things that are customs, habits, ways of acting and thinking, and so on; and finally men in their relations to those still other things that might be accidents and misfortunes such as famine, epidemics, death and so on. (1994a: 209)

In fact, much of this builds on what Foucault had discussed in *Discipline and Punish*. There is, for instance, the same emphasis on the role of the economy in driving political practice (for example, Foucault, 1994a: 207 cites Physiocratic philosopher Francois Quesnay who described good government as 'economic government'), the same concern with the rationalities of liberalism (and in particular how 'freedom' emerges as a strategy of rule), and the same focus on space as a 'field of intervention' in the habits and conduct of populations. These points are drawn from a number of historical examples, including the birth of statistics, the science of demography, and the rapid spread of public health campaigns in the eighteenth and nineteenth centuries, which Foucault saw as evidence that the modern state is increasingly concerned with managing the 'collective health, safety, and aggregate productivity' (Roberts, 2005: 35) of the population (cf. Elden, 2002). Controversially Foucault also seemed to suggest that this problem of 'how to govern the population' coincided with a shift from sovereign power (the power to punish: to seize things and bodies) to 'biopower' ('a power whose highest function was perhaps no longer to kill, but to invest life through and through' (1980: 139)). Foucault was, however, clear that this shift could not be encapsulated in a neat chronology; biopower did not simply replace sovereignty. Indeed, as the final sections in *The History of Sexuality* made clear, the racialized programme of eugenics practised by the Nazis was biopolitical governance practised in its purest form. Under such a regime one killed not people but 'degenerates' in an effort to maintain the 'purity' of the master race. Massacres are thus made to seem 'vital' (Foucault, 1980: 137).

The Spatial Organization of Society

For many geographers Foucault's history of 'governmentality' has proven to be a stimulating way of thinking about the spatial organization of society precisely because it goes beyond his earlier focus on what Rose (1999: 35) calls 'the spaces of enclosure' – hospitals, factories, schools, prisons, etc. – and the 'anatomo-politics' of the disciplinary gaze.

Matthew Hannah (2000), for example, employs Foucault's study of governmentality to shed light on the practice of census-taking in nineteenth-century America. Hannah brilliantly shows how census-taking not only reflected 'the larger cultural context in which it was embedded' (constructing racial categories and identifying 'foreigners', for example) but actually *constituted* the very object (the 'national social body') it was putatively enumerating. Rather than viewing the state as 'authorizing' the national census, Hannah suggests that it would be more accurate to say that the *census produces and authorises the state*. In other words, the 'nation-state' – as a juridical, socio-cultural and political assemblage – emerges as an effect of governmentality.

Other scholars have used Foucault's ideas as a 'toolkit' for thinking about the practice of colonial government. Anthropologists such as Paul Rabinow (1995) and Ann Stoler (1995), historian Gyan Prakash (1999) and political theorist Timothy Mitchell (1991) have each suggested that Europe's colonies are best conceived as 'laboratories of modernity': sites in which the government of society, the discipline of bodies, and the ordering of space was first tested and refined. Mitchell (1991: 35, emphasis added) points out that Foucault's focus on northern Europe 'has tended to obscure the colonising nature of disciplinary power. Yet the Panopticon, the model institution whose geometric order and generalised surveillance serve as a motif for this kind of power, was a *colonial invention*'. 'The panoptic principle', Mitchell goes on to say, 'was devised on Europe's colonial frontier with the Ottoman Empire, and examples of the Panopticon were built for the most part not in northern Europe, but in places like colonial India'. Lacking legitimacy and forced to govern at a distance, colonial powers were often obliged to implement modes of rule that were frowned on and deemed excessive in Europe.

This theme is developed in Stephen Legg's work on 'urban governmentalities' in colonial Delhi. Legg (2007: 24) claims that lingering doubts over 'the ability of colonial populations to support the process on which liberal government relied' forced a reshaping of government, sovereignty and discipline. Fearing that 'uncivilized subjects' were incapable of autonomous development the colonial state was compelled to take on a more directional role in reforming the habits of the population. Disciplinary and sovereign power were expressed through the residential landscapes of New Delhi, where aesthetic, sanitary and policing concerns mapped onto population controls designed to protect privilege and assure order through the partitioning of space (Legg, 2007: 214–215; Scott, 1995).

Nally (2011) explores similar themes in his study of British colonial administration in Ireland. Drawing on Driver's (1993) earlier work, he traces how the administration of the Irish Poor Law (established in 1838, four years after the overhaul of the English Poor Law system) was used to direct changes in the agrarian economy of Ireland and not simply to discipline the lives of individual paupers. Although the author of the Irish Poor Law, George Nicholls (1781–1865), described the new legislation in terms similar to its English cousin – as an instrument of 'enlightened benevolence' and a means 'to revive, or establish, the habit of reliance ... [and] compel them [Irish paupers] to acts of local self-government' – Nicholls also made clear that his wider remit was to implement a 'transition' in the Irish countryside from subsistence farming to agrarian capitalism (Nally, 2011; cf. Foucault, 2007c). Nally argues that this attempt at state-led social engineering (Scott, 1998) marks a shift from disciplinary power – targeting the body of the pauper – to a mode of biopolitics that targeted the Irish social body much more generally. The striking thing about these reforms was the attempt to align colonial paternalism to a new politics of racial improvement (cf. Duffield and Hewitt, 2009; Venn, 2009). Nally shows how this desire to 'help the poor' (see also Chapter 3) was gradually transformed into a violent programme of clearances and ejectment during the Great Irish Famine. As Tania Li (2007: 68) has discussed elsewhere, the colonial 'will to improve' was 'difficult to reconcile with high mortality, although not it seems impossible'. In Ireland the hope of promoting a 'salutary revolution in the habits of a nation' (Trevelyan, cited in Nally, 2011: 209) led many elites and officials to embrace famine as a shortcut to modernity.

This synthesis between 'making live' and 'letting die' was for Foucault (1980) the very hallmark of modern state racism (cf. Foucault, 2003). According to Cameroonian scholar Achille Mbembe (2003: 23–24) the attempt 'to attribute rational objectives to the very act of killing' was most commonly expressed *outside* the colonial metropole: 'the colonies are the location par excellence where controls and guarantees of judicial order can be suspended – the zone where the violence of the state of exception is deemed to operate in the service of "civilization"'. In the colonies killing was routinely sanctioned (and sanitized) as a means of eradicating deviancy, enforcing industry and preserving the good order of society. Killing, or exposing populations to likely fatality, was made to seem vital.

This tension between liberal democratic values and colonial governmentalities is productively explored in James Duncan's study of the nineteenth-century plantation economy in Ceylon. Duncan (2007: 7) brilliantly traces the arbitration processes 'that decided the moral acceptability of death rates' on the coffee plantations. The planter's quest to discipline estate labour to ensure maximum productivity met the problem of having to work with and on populations exhibiting 'differentiated capacities and potentialities' (Duncan, 2007: 21). Duncan (2007: 190) argues that the need to discipline workers for the rigours of commodity production led planters to resist 'ideas of humane governmentality emanating from Britain' and adopt instead a form of 'authoritarian biopower which at best sought to merely make survive'. Indeed for some officials the collection of taxes, which left locals battling against starvation and eviction, were a *necessary* imposition. If these fiscal burdens were attenuated the state would lose a powerful means to encourage native exertion (Duncan, 2007: 182).

This brand of colonial stewardship was strikingly similar to strategies of rule adopted by the British in the Caribbean. With the passing of the Slavery Abolition Act in 1833, plantation owners faced the problem of how to transform slave labourers – considered racially inferior and predisposed to idleness – into self-motivated 'free' workers. According to Thomas Holt the

> 'problem of freedom' ... could be addressed only by thoroughly reforming the ex-slaves' culture so as to make them receptive to the discipline of free labour. Specifically, a social environment needed to be created that would make them disconnected with a mere subsistence-level existence; their material aspirations needed to be gradually expanded. This process would instil the internal discipline to make them a reliable working class. (1992: xxii)

'A common theme running through racist thought', continues Holt (1992: 307), 'was that "the natives" had no inner controls; thus the need for external controllers' (see also Chapter 8). The decision to free British slaves was thus the 'moment when classical liberalism's internal contradictions stand exposed' (Holt, 1992: xix) – which is to say, the moment when 'freedom' is revealed as means for governing the conduct and aspirations of former slaves. It bears repeating that the older model of sovereign power – with its emphasis on violence and physical subjugation – was never wholly renounced as the state embraced new and 'milder' forms of governmentality based on ordering conduct to ensure maximum productivity. The 'dark side' of modernity that Foucault addressed was precisely the potential for 'order' to harbour within itself the seeds of violence. As Martinique scholar Frantz Fanon (1967: 29) remarked, 'if philosophy and intelligence are invoked to proclaim the equality of men, they have also been employed to justify the extermination of men'.

Conclusion

Foucault's analysis of the administration, control and direction of populations has opened up fresh ways for geographers to think about a wide range of issue including public health (Brown, 2009), urban design (Yeoh, 1996), demography (Legg, 2005), food provisioning (Nally, 2008), war (Gregory, 2004; Tyner, 2009), crime (Herbert, 1996), race (Crampton, 2007), epidemiology (Kearns, 2007), prostitution (Howell, 2009) and mapping (Rose-Redwood 2006; Crampton, 2007). Some of this research expands on themes addressed by Foucault in some detail, but more commonly this work develops subjects that are only briefly mentioned in his lectures and seminars or not explored at all. Having once described his work as taking place 'between unfinished abutments and anticipatory strings of dots', it seems fitting that geographers would help to continue a conversation that clearly never finished. 'My books aren't treatises in philosophy or studies of history', Foucault (1994b: 222–223) went on to clarify, 'they are philosophical fragments put to work in a historical field of problems.' For this reason Philip Howell (2007: 292) is surely right to insist that 'the best work in geography has been done *with* and *after* Foucault rather than simply *for* or *against* him'.

KEY POINTS

- Governmentality describes the tactics and strategies for rendering society more governable.
- Governmentality builds on Foucault's earlier discussion of disciplinary power, which he defined as the production of 'useful' and 'docile' bodies.
- Historical geographers have used Foucault's ideas to show how power is directed toward the spatial organization of society.
- Governmentality can develop in 'benign' and 'authoritarian' forms and historical geographers have sought to trace how 'liberal' strategies of rule can produce decidedly violent outcomes.

References

Bentham, J. (1843) 'Panopticon; or the Inspection house', in J. Bowring (ed.) *The Works of Jeremy Bentham Published under the Superintendence of his Executor John Bowring, Volume 4*. Edinburgh: William Tait, pp. 37–172.

Brown, M. (2009) 'Public health as urban politics, urban geography: venereal biopower in Seattle 1943–1983', *Urban Geography*, 30(1): 1–29.

Crampton, J. (2007) The biopolitical justification for geosurveillance', *Geographical Review*, 97(3): 389–403.

Driver, F. (1993) *Power and Pauperism: The Workhouse System, 1834–1884*. Cambridge: Cambridge University Press.

Duffield, M. and Hewitt, V. (eds) (2009) *Empire, Development and Colonialism: The Past in the Present*. Woodbridge: James Currey.

Duncan, J.S. (2007) *In the Shadows of the Tropics: Climate, Race and Biopower in Nineteenth Century Ceylon*. Aldershot: Ashgate.

Elden, S. (2002) 'The war of race and the Constitution of the State: Foucault's "Il faut défendre la société" and the politics of calculation', *Boundary 2*, 29(1): 125–151.

Fanon, F. (1967) *Black Skin, White Mask* (translated by Charles Lam Markmann) New York: Grove Press.

Foucault, M. (1979) *Discipline and Punish: The Birth of the Prison*. New York: Vintage.

Foucault, M. (1980) *The History of Sexuality. Vol. 1: An Introduction*. New York: Vintage.

Foucault, M. (1984) 'Space, knowledge and power', in P. Rabinow (ed.) *The Foucault Reader*. Harmondsworth: Penguin, pp. 239–256.

Foucault, M. (1994a) 'Governmentality', in J.D. Faubion (ed.) *Essential Works of Foucault 1954–1984, Volume Three: Power*. New York: The New York Press, pp. 201–222.

Foucault, M. (1994b) 'Questions of method', in J.D. Faubion (ed.) *Essential Works of Foucault 1954–1984 Volume Three: Power*. New York: The New York Press, pp. 223–238.

Foucault, M. (2003) *Society Must be Defended: Lectures at the Collège De France 1975–1976*. New York: Picador.

Foucault, M. (2007a) 'Questions on geography', in J. Crampton and S. Elden (eds) *Space Knowledge and Power*. Farnham: Ashgate, pp. 173–184.

Foucault, M. (2007b) 'The incorporation of the hospital in modern technology', in J. Crampton and S. Elden (eds) *Space Knowledge and Power*. Farnham: Ashgate, pp. 141–152.

Foucault, M. (2007c) *Security, Territory, Population: Lectures at the Collège de France 1977–1978*. New York: Palgrave Macmillan.

Gregory, D. (2004) *The Colonial Present: Afghanistan, Palestine, Iraq*. Oxford: Blackwell.

Hannah, M. (2000) *Governmentality and the Mastery of Territory in Nineteenth-Century America*. Cambridge: Cambridge University Press.

Herbert, S. (1996), 'The geopolitics of the police: Foucault, disciplinary power and the tactics of the Los Angeles Police Department', *Political Geography*, 15(1): 47–59

Holt, T.C. (1992) *The Problem of Freedom: Race, Labor, and Politics in Jamaica and Britain, 1832–1938*. Baltimore, MD: Johns Hopkins University Press.

Howell, P. (2007) 'Foucault, sexuality, geography', in J. Crampton and S. Elden (eds) *Space, Knowledge and Power: Foucault and Geography*. Farnham: Ashgate.

Howell, P. (2009) *Geographies of Regulation: Policing Prostitution in Nineteenth-Century Britain and the Empire*. Cambridge: Cambridge University Press.

Kearns, G. (2007) 'The history of medical geography after Foucault', in J. Crampton and S. Elden (eds) *Space, Knowledge and Power: Foucault and Geography*. Farnham: Ashgate, pp. 205–222.

Legg, S. (2005) 'Foucault's population geographies: classifications, biopolitics and governmental spaces', *Population, Space and Place* 11(3): 137–156.

Legg, S. (2007) *Spaces of Colonialism: Delhi's Urban Governmentalities*. London: Blackwell.

Li, T.M. (2007) *The Will to Improve: Governmentality, Development, and the Practice of Politics*. Durham, NC: Duke University Press.

Losurdo, D. (2011) *Liberalism: A Counter-history*. London: Verso.

MacKenzie, D. (1984) 'Marx and the machine', *Technology and Culture*, 25(3): 473–502.

Mayhew, R.J. (2009) 'Historical geography 2007–2008: Foucault's avatars – still in (the) Driver's seat', *Progress in Human Geography*, 33(3): 1–11.

Mbembe, A. (2003) 'Necropolitics', *Public Culture*, 15(1): 11–40.

Mitchell, T. (1991) *Colonising Egypt*. Berkeley: University of California Press.

Nally, D. (2008) '"That coming storm": the Irish Poor Law, colonial biopolitics, and the Great Famine', *Annals of the Association of American Geographers*, 98(3): 714–741.

Nally, D. (2010) 'The biopolitics of food provisioning', *Transactions of the Institute of British Geographers*, 36(1): 37–53.

Nally, D. (2011) *Human Encumbrances: Political Violence and the Great Irish Famine*. Notre Dame, IN: University of Notre Dame Press.

Philo, C. (2004) *A Geographical History of Institutional Provision for the Insane from Medieval Times to the 1860s in England and Wales: 'The Space Reserved for Insanity'*. Lampeter: Edwin Mellen Press.

Prakash, G. (1999) *Another Reason: Science and the Imagination of Modern India*. Princeton, NJ: Princeton University Press.

Rabinow, P. (1995) *French Modern: Norms and Forms of the Social Environment*. Chicago: University of Chicago Press.

Roberts, W. (2005) 'Sovereignty, biopower and the state of exception: Agamben, Butler and indefinite detention', *Journal for the Arts, Sciences, and Technology*, 3(1): 33–40.

Rose-Redwood, R. (2006) 'Governmentality, geography, and the geo-coded world', *Progress in Human Geography*, 30(4): 469–486.

Rose, N. (1999) *Powers of Freedom: Reframing Political Thought*. Cambridge: Cambridge University Press.

Scott, D. (1995) 'Colonial governmentality', *Social Text*, 43: 191–220.

Scott, J. (1998) *Seeing Like a State: How Certain Schemes to Improve the Human Condition Have Failed*. New Haven, CT: Yale University Press.

Stoler, A. (1995) *Race and the Education of Desire: Foucault's History of Sexuality and the Colonial Order of Things*. Durham, NC: Duke University Press.

Tyner, J.A. (2009) *War, Violence, and Population: Making the Body Count*. New York: Guilford Press.
Venn, C. (2009) 'Neoliberal political economy, biopolitics and colonialism: a transcolonial genealogy of inequality', *Theory, Culture & Society*, 26(6): 206–233.
Yeoh, B. (1996) *Contesting Space: Power Relations and the Urban Built Environment in Colonial Singapore*. Oxford: Oxford University Press.

FURTHER READING

Crampton, J. and Elden, S. (eds) (2007) *Space, Knowledge and Power: Foucault and Geography*. Farnham: Ashgate.
Duncan, J.S (2007) *In the Shadows of the Tropics: Climate, Race and Biopower in Nineteenth Century Ceylon*. Aldershot: Ashgate.
Foucault, M. (1994a) 'Governmentality', in J.D. Faubion (ed.) *Essential Works of Foucault 1954–1984, Volume Three: Power*. New York: The New York Press, pp. 201–222.
Legg, S. (2007) *Spaces of Colonialism: Delhi's Urban Governmentalities*. London: Blackwell.
Nally, D. (2011) *Human Encumbrances: Political Violence and the Great Irish Famine*. Notre Dame, IN: University of Notre Dame Press.

21 NATURE–CULTURE

David Nally

Introduction

Nature has been aptly described as one of the most 'potent and ambiguous words' in the English language (Haraway, 1991: 1; Foster et al., 2010: 32). Certainly meanings of nature abound. For some people nature describes what is seemingly most sacred about human life; it designates our 'core' or 'essence', the very stuff that defines our inner lives and marks us out as individuals (see Chapter 9). In the world-view of others, nature is primarily a 'pristine' environment, a destination one visits precisely in order to seek refuge from the hustle and bustle of everyday life; here 'nature' is understood as the obverse of 'society'. For the smallholder farmer, nature is bounty and satiation – the means to a full stomach. Through loudspeakers at environmental rallies activists present nature as something to be 'saved' (or more nostalgically as that which is 'lost' to progress), while in company boardrooms nature is something to be 'managed' under the aegis of 'corporate social responsibility'. Early colonial settlers thought of nature as an open frontier – 'virgin' territory to be tamed and domesticated – whereas for Malthus and his apostles nature presented an irrevocable limit or 'check' on human profligacy (see Chapter 9). In the eyes of the first industrialists nature was an inert input, a 'raw material' fuelling economic growth, whereas for many ecologists nature, reconceptualized as a 'biosphere', is a dynamic system that actively regulates, rather than simply supports, life on planet earth.

Evidently the nature we see and the ideals we envisage – including the 'future natures' we aspire to – are heavily filtered (that is to say, mediated) – by our cultural beliefs and sensibilities. The human presence in nature is undeniable, leading some to conclude that nature can no longer be judged to occupy a 'discrete ontological space' (Braun and

Castree, 1998; Hinchliffe, 2007), distinct and separate from the sphere of human activity (cf. Haslanger, 1995). In the vivid words of Castree and Braun (2001), social relations extend 'all the way down' so that there are no clear metaphysical distinctions to be drawn between natural and social objects. These ideas have gained wide appeal in Geography – a discipline that desperately, and quite reasonably, wants to distance itself from its historical association with environmental determinism (see Chapter 8). The old view that human beings are 'creatures of nature' has rightly been discredited; indeed, nowadays it is more common to hear geographers – and other social scientists – ask whether there is even a nature 'out there' to be 'defended and preserved' (Braun and Castree, 1998: xi). This position is neatly summarized by Donna Haraway (1991), who argues that nature does not in any sense precede its social construction. All that is solid seemingly melts into air.

Historicizing Nature

Some of the most striking and original work on the concept of nature stems from geographers working in the tradition of historical materialism. At the head of the queue is Neil Smith's foundational study *Uneven Development: Nature, Capital and the Production of Space*. Smith's task was to describe how nature emerges in relation to capitalist development. This move to *historicize* nature – if it can be phrased this way – is best captured in the distinction drawn between 'first' and 'second' nature (cf. Schmidt, 1971). The former concept refers to the world 'outside' and beyond human activity, while the latter describes the transformations pressed on the material world by human beings. According to Smith 'second nature' emerges with the historical development of exchange economies. 'With production for exchange', Smith (2008: 65) explains, 'the production of nature takes place on an extended scale. Human beings not only produce the immediate context of their existence, but produce the entire societal nature of their existence'. For Smith 'societal' or 'second' nature emerges from a forked process. First, exchange economies generate new complex divisions of labour that differentiate human beings according to class, sex, gender, and so on. Second, capitalist production stitches together 'previously, isolated localized groups of people' and in the process constitutes wholly novel social formations dominated by the market, state, private property and the bourgeois family (Smith, 2008: 65). As Smith concludes: 'Society as such, clearly distinguishable from nature emerges ... Second nature is produced out of first nature.'

Smith's analysis is a powerful critique of the production of social nature under historical capitalism. The reification of the market system and the advance of capitalism as a globalizing force (see Chapter 19) mean that first nature is reconstituted as social matter fashioned by human relations. Importantly, Smith (2008: 68) is clear that the production of this 'second nature' does not mean that matter 'cease[s] to be nature in the sense that ... [it is] now immune from non-human forces and processes'. This, in other words, is not a hyper-constructionist approach to nature – the view that 'a fish is only a fish if it is socially classified as one', as Keith Tester (cited in Foster, et al., 2010: 34) argues – but rather a dialectical account that stresses how humans are forcing 'qualitative' and 'quantitative' changes on the world that in turn shapes and re-shapes us as human beings. '[T]he relation with nature', Smith reminds us, 'develops *along with* the development of social relations, and in so far as the latter are contradictory, so too is the relation with nature' (2008: 68, emphasis added).

The historicization of nature sets up Smith's larger theory of uneven geographical development. For Smith overproduction and underproduction, accumulation and dispossession, poverty and wealth, human flourishing (for some) and social debasement (for others) are two sides of the one coin. Smith's thesis is a 'contrapuntal' (Said, 2000: 186) reading of socio-economic history and a re-positioning of the historical geography of nature (Simmons, 1998). Geographers and social scientists working in cognate fields have taken up this method and employed it usefully in other contexts (see also Chapter 10). For example, scholars associated with the 'New Western History' (so termed by historian Patricia Nelson Limerick) have re-examined American frontier history with an eye to critically recapturing occluded histories and geographies (cf. Zinn, 1980) and undermining 'the central myth of the frontier as the place where civilization overcame savagery' (Kearns, 1998: 377). 'Europeans did not find wilderness here [in the Americas]', writes historian Francis Jennings (cited in Wright, 2005: 113), 'they made one.'

Without romanticizing the First Nations' cultures that were obliterated as the American frontier expanded, scholars such as William Cronon and Donald Worster have sought to show how the commodification of nature redefines resources in such a way as to undermine local and independent farming cultures. In other words, the production of 'second nature' from 'first nature' was a vehicle for the social control of subaltern peoples. As Kearns (1998: 389, 396) puts it in his fine review of the contributions of the New Western History: 'The drive to dominate nature entails the domination of people ... Native agricultural systems did not fall apart from within in some Malthusian inspired disaster; rather settlers waged an explicit war against indigenous subsistence cultures.'

In the colonies, however, the expropriation and reconstitution of 'first nature' was far more rapid and force was regularly used as a surrogate to statecraft. For example in Karamoja, Uganda, according to political scientist Mahmood Mamdani (1982: 68), British colonialism began with the forcible acquisition of natural capital leaving local people bereft of the means of production and thrust back upon precarious modes of pastoral cultivation. In Nigeria (Watts, 1983), Ireland (Nally, 2011b), and India (Davis, 2001) the commodification of nature led to hugely destructive famines that eliminated whole communities and ways of living. Put simply, the production of 'second nature' by colonial authorities pushed marginalised groups into an often-deadly relationship with 'first nature'. Thus what are frequently termed 'natural disasters' are in fact the result of the forceful introduction of an entirely new – and for many an increasingly precarious – relationship with nature (Pelling, 2001; Smith, 2006).

A recent study of the plantation economy confirms the degree to which colonial rule was designed to manage 'first nature' for the benefit of the metropole. Focusing on the coffee plantations in colonial Ceylon geographer James Duncan (2007: 35, 40) highlights the efforts to harness human nature (the daily disciplining of bodies to make them receptive to producing for the market) and the environment (engineering entirely new 'monocultural' ecologies) in order to maximise profits. Duncan's research shows that plantation economies birthed not simply a new political relationship between the colonies and the metropole, but an *entirely new social relationship with nature.*

It is crucial to recognize the degree to which this brand of 'biological imperialism' (Crosby, 1986) was central to the 'take off' of European states. European powers benefited enormously from their ecological experiments, with some scholars even arguing that the 'ghost acres' seized through colonial dispossession (the extraterritorial land, that is, which increased the home nation's carrying capacity) helped stave off demographic pressures that otherwise would have attenuated European growth. Importantly colonial trade brought new root crops – especially white potato, cassava and sweet potatoes – to the 'Old World'. Europeans also introduced maize, groundnuts, tomatoes, cassava, sweet potatoes and tobacco to Africa (where they became established in slash-and-burn agricultural cycles and 'cash crop' production patterns) while tomatoes, chilli peppers and sweet potatoes were introduced to China and are sometimes credited with sustaining the rise in the Chinese population from the sixteenth century (Kiple, 2007). Furthermore, these colonial 'ghost acres' arguably enabled European nations such as Britain to urbanize, but also to develop a fully industrial economy.

Significantly, these forced food transfers meant lifting single cultigens out of their native production, processing and consumption contexts (Crossgrove et al., 1990: 228). The transfer process generally did not include complementary species resulting in artificial mono-cultural environments that were nutritionally insufficient and often significantly more vulnerable to attacks by pathogens and insects. The so-called 'Irish potato famine' of the 1840s is a well-known example of the consequences of over-dependence on a single root crop. The reconstitution of nature undertaken during the 'Great Columbian Exchange' (Crosby, 1972; Grove, 1995) also made pellagra an endemic disease in Africa, Mexico, Indonesia and China, as poor populations turned to maize as a cheap source of calories. Thus the historical quest to bring to heel 'first nature' often left local populations alienated from the means of production, more vulnerable to exogenous shocks, and more susceptible to malnutrition and nutrition-related diseases. Conversely the production of 'second nature' developed alongside a colonizing discourse that reconstructed 'the tropics' as the environmental Other of the West (Clayton, 2013). Indeed, the conversion of 'first nature' into 'second nature' was viewed as tangible proof of European superiority, while the crippling underdevelopment of the global South was rationalized as a 'natural order'. The so-called Third World was thronged with peoples who were seemingly incapable of securing the natural advantages of the soil. In this way 'wild nature' was symbolically conflated with an 'untamed' population (Gregory, 2001)

Molecularizing Nature

More recently social scientists have turned their attention to developments in the life sciences – particularly in genomics – where the employment of recombinant DNA technologies is promising to revolutionize not only how we do science, but our very understanding of life itself (see also Chapter 17). In the field of health care, scientists claim the ability to create cells that can be 'silenced' or turned off – a manipulation that some believe will lead, for instance, to a permanent cure for cancer. A recent paper published in the peer-review journal *Ethics, Policy and Environment* argued that 'human engineering' – now technically possible – might even be a viable solution to global climate change. Authors Liao et al. (2012) suggest that genetic technologies could be used to induce intolerance to red meat, thereby limiting the greenhouse gas emission associated with modern livestock farming. Using techniques such as pre-implantation genetic diagnosis humans might also 'select' smaller embryos generating smaller children who will require less energy. Even more outlandishly the authors discuss methods to increase the cognitive capacity of human subjects as a means to diminish birth rates – their hypothesis being that smarter people have fewer children. Anticipating a sharp reaction to their research the authors conclude that their arguments might even be 'liberty enhancing'. If these proposals were implemented, they argue, China could dispense with its controversial 'one child' policy; human beings would be smarter and presumably more capable of great things; the reduction in red-meat consumption would dampen pressure on overburdened health-care services and minimize the suffering of non-humans; there would be less of a need to punitively tax undesirable human behaviour, and so on (Liao et al., 2012).

Ethical questions abound, but for the purposes of this chapter it is interesting for us to think about what such apparently far-fetched proposals mean for our understanding of 'human nature'. To some, such drastic 'eugenic' manipulation of human 'nature' replicates or intensifies what humans have accomplished with non-human animal natures since the domestication of dogs, goats, sheep and other animals over 10,000 years ago, and with the dawn of 'scientific breeding' little more than 200 years ago. The application of such techniques raises the spectre of the animalization of humans, and the wholesale dismantling of the distinction between humans and other animals that the Italian philosopher Giorgio Agamben has referred to in terms of the workings of an 'anthropological machine'. For Agamben, therefore, 'the decisive political conflict, which governs every other conflict, is that between the animality and the humanity of man' (2004: 80). Other prominent philosophers who have taken up this 'biopolitical' question, critically developing Foucault's post-humanist stance beyond its original formulation, include Gilles Deleuze, Jacques Derrida,

Donna Haraway and Roberto Esposito, among many others. Their theorization of biopolitics (see Chapter 20) has vital significance for how we think and treat non-human animals, but also, of course, how we think and how we treat human beings, including the differently abled, non-Western peoples, children, women, the poor, and others who have typically been seen as 'closer' to animals than the white, male, adult and fully 'civilized' Western subject. Summing up the implications of Foucault's biopolitics, Esposito (2008: 29), suggests that 'a definable and identifiable human nature' no longer exists, at least 'independent from the meanings that culture and therefore history have, over the course of time, imprinted on it'. We can ask, in short: what is 'human nature' if life itself becomes 'infinitely malleable' (Rose, 2007: 1) and the 'self-instrumentalization of the species' is made technically possible (Habermas, 2003)?

Genetic science is forcing us to radically rethink our understanding of nature. In 2010 several scientists working at the J. Craig Venter Institute announced that they had created for the first time an entirely new form of artificial life. An article that followed in *The Economist*, entitled 'Genesis Redux', spelt out the significance of this breakthrough: '[The scientists] have done for real what Mary Shelley merely imagined ... The result is the first creature since the beginning of creatures that has no ancestor' (*Economist*, 2010: n.p.) For some scholars such developments herald the dawn of a 'third nature' marked by a 'post-ontological' appreciation of life (Fukuyama, 2003; Habermas, 2003) and a shift to a 'world after nature' (McKibben, 2006). Such views are immensely challenging for those who wish to hold onto a transcendental understanding of nature. In their view, human life is sacred and altering nature is tantamount to 'playing God' (see Chapter 9).

To return to Neil Smith's terminology for a moment, there is little doubt that 'nature' has long been shaped by human activity. Of course the difference now is that scientific discoveries are reinventing the parameters of what it means to 'shape' nature. Again the 'qualitative' and not merely the 'quantitative' dimensions of these changes are important (Smith, 2008: 77). As we have seen, Smith's account underscored the development of capital at a world scale and the transformations wrought on the material world. The capitalist 'appropriation of nature', he said, 'is accomplished not for the fulfilment of needs in general, but for the fulfilment of one need in particular: profit' (Smith, 2008: 78). For geographer Sian Sullivan (2013: 199) contemporary conservation initiatives perfectly illustrate how global 'nonhuman natures and nature dynamics' are being remade to generate greater profits. Far from signalling a developmental crisis of capitalism, Sullivan argues, environmental problems are fast becoming an 'accumulation frontier' as 'conserved nature' is repackaged in monetary and tradable terms (2013: 200; see also Baldwin, 2009). The irony here is that environmental problems – many of of them brought about by capitalism – bring with them opportunities to further insert nature into regimes of accumulation.

In a similar vein a number of geographers (Fitzsimmons and Goodman, 1998; Castree, 2001; Nally, 2011a) have explored how new kinds of 'agricultural nature' (Braun and Castree, 1998: 10; Powell, 2000) are being forged to facilitate capital accumulation. They argue that genetically modified organisms (GMOs), governed by intellectual property rights (IPR), pave the way for the wholesale privatization of nature – and as eco-feminist Vandana Shiva (2010: 234) reminds us, the Latin route of the word 'private' means 'to deprive'. Crops that once produced their own seed can now be rendered sterile – effectively turning a renewable 'first' nature into a non-renewable second (third?) nature – while biomass formerly used for food, fodder and shelter can be broken down and reconstituted as biofuel sold on to affluent consumers. The United Nations recently announced that biofuels are now the fastest growing segment of the world agricultural market.

The ability to retrofit our volatile hydrocarbon economy to accommodate fuel derived from biomass has also, undoubtedly, accelerated a global rush to purchase land in the global South (so-called land grabs). Thus the meteoric emergence of the biofuels industry comes with severe costs for the world's poor. Filling the tank of a sport utility vehicle, for example, uses approximately 450 pounds of corn – enough food to feed one person for an entire year. Seen this way the production of 'agricultural natures' designed to enhance the 'energy security' of relatively affluent countries in Europe or North America can compromise the food sovereignty of peoples in poorer parts of the world (cf. Nally and Kearns, 2011).

Above we saw how the severing of 'organic' ties between peoples and places, and the invention of a wholly 'new metabolism with nature' (Wood, 2000: 39), was a central feature of colonial plantations. But whereas colonial agriculture depended on a monopoly of trade and experimental forms of labour control, modern agribusiness arguably rests on the *monopolization of life and living resources* (Shiva, 2000: 3), a feat made possible by new agro-biotechnologies and the increased capacity 'to control, manage, engineer, reshape, and modulate the very vital capacities of human beings as living creatures' (Rose, 2007: 3). The tightening relationship between the biosciences and capitalist agribusiness has led to biological interventions that have, amongst other things, accelerated the commodification of nature in ways that were not previously possible; Marx was largely right, then, when he described capitalism as a machine for demolishing limits (Calasso, 1994: 237). This massive realignment of nature – a 'qualitative' and 'quantitative' change unprecedented since the 'Great Columbian exchange' – proves the ever-dynamic complexion of the social engagement with nature.

Nature in the Anthropocene

In one of his final books before he died Michel Foucault wrote, 'For millennia, man [sic] remained what he was for Aristotle: a living animal with the additional capacity for political existence; modern man is an animal whose politics calls his existence as a living being into question' (1980: 143). These are increasingly prescient words. So extensive and intensive has been the human appropriation of nature that many scientists now claim that humanity has entered a new historical era appropriately termed the Anthropocene (meaning 'new human'). Although humans as a species have inhabited the planet's surface for 'less than 1% of 1% of its history' the evidence that human-induced changes are reshaping the planet on a 'geological scale' – and crucially 'far-faster-than-geological speed' (*Economist*, 2011) – is increasingly difficult to deny. The authors of *The Ecological Rift* (Foster et al., 2010) cite ocean acidification, biodiversity loss, stratospheric ozone depletion, atmospheric aerosol loading, chemical pollution, changes to the nitrogen and phosphorous cycles, and the depletion of global fresh water stocks as evidence that ecological limits are fast being breached. The absorptive capacity of the planet is being challenged to such a degree that the very existence of humans as a species is now also under threat (Dalby, 2013).

If it is accepted that humans are now a *force of nature* in the geological sense then according to historian Dipesh Chakrabarty (2009), there is a clear need to develop a new kind of 'species thinking' in the social sciences. As Chakrabarty (2009: 221) puts it:

> The anxiety global warming gives rise to is reminiscent of the days when many feared a global nuclear war. But there is a very important difference. A nuclear war would have been a conscious decision on the part of the powers that be. Climate change is an unintended consequence of human actions and shows, only through scientific analysis, the effects of our actions as a species. Species may indeed be the name of a placeholder for an emergent, new universal history of humans that flashes up in the moment of the danger that is climate change.

While human geographers – from all subfields – have made strong contributions to the study of 'social nature' it is quite remarkable that there has been such audible silence on the issue of anthropogenic climate change. Perhaps this is because climate change is a 'low intensity' problem – a form of 'slow violence' to use Rob Nixon's (2011) term – that either goes unnoticed or we quickly become inured to it (Buell, 2004). In the words of Simon Dalby (2007a: 160), one of the few human geographers working on this pressing topic (but see also Gibson Graham and Roelvink, 2009), we are effectively 'constructing the context for our lives at the very biggest of scales: the planetary biosphere itself' and, importantly, without full awareness of the consequences. There are profound challenges here for geographers and social scientists. 'Species thinking' of the kind of that Chakrabarty (2009) advocates, means grasping non-linear forms of causality, including new forms of 'catastrophic convergence': not simply several disasters happening simultaneously ('one problem atop another') but the potential for 'problems to compound and amplify one another' (Parenti, 2011: 7). Ecologists frequently speak of 'tipping points' and 'feedback loops', but what, asks social psychologist Harald Welzer, are the equivalents for describing *social action* in the Anthropocene? It is important to have answers to such questions – or least sound approximations – in order to better understand how humanity is most likely to react when faced with the kinds of convergent crises that are predicted (Dalby, forthcoming).

Conclusion

For Welzer, 'species thinking' also requires new reflective forms capable of grasping complex chains of agency and discordant sequences of action. 'When chains of action extend in space/time', writes Welzer, 'causality also drifts: social action, is not played out in a chain a-b-c-d-e, or as a sequence of action and reaction, but rather as a *development of relations*' (2012: 87, emphasis added). In other words, the issue of anthropogenic climate change requires us to develop modes of reasoning that accord non-human actors 'careful symmetry' (Robbins and Moore, 2012: 15) with human actors whilst also taking due cognisance of the fact that the places likely to be hardest hit by climatic perturbations are unlikely to be the places associated with the actions and processes that caused the problems in the first place. Geographers have arguably made great strides in analysing such 'more-than-human' or 'hybrid' geographies (Whatmore, 2002); and, to a degree, they have also advanced new relational modes of thinking (Jones, 2009) that challenge outmoded, mechanical accounts of causality.

In getting to grips with anthropogenic climate change geographers might also draw lessons from the historical analysis of technological change. Indeed, the ability to transfer the costs of climate change echoes the ongoing capacity to relocate the deleterious effects of rapid growth and industrialization. Otto Ullrich's (2010: 318) insightful reflections on this are worth citing at length:

> The capacity to transfer costs makes it possible for modern technology to appear in a mystified form. It tricks the senses as to its performance capacities. The costs are usually transferred and scattered over very considerable times and spaces. The spatial and temporal horizon of our perception is, however, significantly nearer. What we know of measured pollution levels, and the costs in the future or in distant areas, remains abstract to us and too far removed from currently perceived realities. It touches none, or too few, of the feelings and thoughts that determine behaviour here and now. Who can imagine a 300,000-year, radioactive half-life in concrete form? How much does the knowledge of a hole in the ozone layer count for against the utility advantage, impressed upon our senses right now, of instantly available cool drinks from the fridge or the comfortable transportation offered by a high-performance private automobile? The temporal, spatial and personal separation of utilities and costs – the separation of an act committed now from the suffering that ensues, or the non-intersection between advantages that are privately consumable and disadvantages that have to be borne collectively – is an exceedingly seductive characteristic of modern scientific technologies.

If the future is to be a promise rather than a threat, it is surely urgent to develop modes of reflection that can stitch together social problems that at first blush appear discrete and non-synchronous.

But isn't this topic far removed from the concerns of historical geographers? Not necessarily: the more politics becomes future-oriented the more imperative it is that we retain a rigorous historical awareness. For

Søren Kierkegaard was certainly correct when he famously noted that while life is lived forward it can only ever be understood backward. This backward-looking sensibility is crucial to thinking about ever-changing nature–culture relations and to assessing the future as it flashes up moments of the danger.

KEY POINTS

- While the distinction drawn between 'first nature' – signifying the world 'outside' and beyond human activity – and 'second nature' – signifying the transformations pressed on the world by humans – is a helpful heuristic device, many geographers now argue that the human presence in nature is indisputable. For them social relations extend 'all the way down' and there are no clear distinctions to be drawn between 'natural' and 'social' objects.
- The historicization of nature, particularly as theorized by Neil Smith, opens up fresh ways to think about how historical capitalism has forced 'quantitative' and 'qualitative' changes on the material world.
- The commodification of nature has accelerated in certain historical periods. The production of agrarian natures following the 'discovery' of the New World created an entirely new 'metabolism' with nature, while the present 'bio-revolution' promises ever more radical ways to modulate and engineer nature. In many cases the overriding objective is to maximize profit.
- The advancing climate crisis represents another precipitous shift in our social relations with nature. Such has been the human impact on the material world that many scientists now claim that we have entered a new historical era, the Anthropocene. Fresh modes of relational thinking will be required to take stock of these potentially life-altering changes.

References

Agamben, G. (2004) *The Open: Man and Animal*. Stanford, CA: Stanford University Press.

Baldwin, A. (2009) 'Carbon nullius and racial rule: race, nature and the cultural politics of forest carbon in Canada', *Antipode*, 41(2): 231–255.

Braun, B. and Castree, N. (eds) (1998) *Remaking Reality: Nature at the Millennium*. London: Routledge.

Buell, F. (2004) *From Apocalypse to Way of Life: Environmentalism in the American Century*. London: Routledge.

Calasso, R. (1994) *The Ruin of Kasch* [translated by William Weaver and Stephen Sartarelli]. Harvard: Carcanet Press.

Castree, N. (2001) 'Marxism, capitalism and the production of nature', in N. Castree, and B. Braun (eds) *Social Nature: Theory, Practice, and Politics*. Oxford: Blackwell, pp. 189–207.

Castree, N. and Braun, B. (eds) (2001) *Social Nature: Theory, Practice, and Politics*. Oxford: Blackwell.

Chakrabarty, D. (2009) 'The climate of history: four theses', *Critical Inquiry*, 35: 197–222.

Clayton, D. (2013) 'Militant tropicality: war, revolution and the reconfiguration of "the tropics"', *Transactions of the Institute of British Geographers*, 38(1): 180–192.

Crosby, A.W. (1972) *The Columbian Exchange: Biological and Cultural Consequences of 1492*. Westport, CT: Greenwood Press.

Crosby, A.W. (1986) *Ecological Imperialism: The Biological Expansion of Europe, 900–1900*. Cambridge: Cambridge University Press.

Crossgrove, W., Egilman, D., Heywood, P., Kasperson, J.X., Messer, E. and Wessen, A. (1990) 'Colonialism, international trade, and the nation-state', in L. Newman (ed.) *Hunger in History: Food Shortage, Poverty and Deprivation*. Oxford: Basil Blackwell, pp. 215–240.

Dalby, S. (2007a) 'Ecology, security, and change in the Anthropocene', *Brown Journal of World Affairs*, 13(2): 155–164.

Dalby, S. (2007b) 'Anthropocene geopolitics: globalisation, empire, environment and critique', *Geography Compass*, 1(1): 103–118.

Dalby, S. (2013) 'Biopolitics and climate security in the Anthropocene', *Geoforum*, 49: 184–192.

Dalby, S. (forthcoming) 'Rethinking geopolitics: climate security in the Anthropocene', *Global Policy*.

Davis, M. (2001) *Late Victorian Holocausts: El Niño Famines and the Making of the Third World*. London: Verso.

Drayton, R. (2002) 'The collaboration of labour: slaves, empires, and globalizations in the Atlantic World', in A.G. Hopkins (ed.) *Globalization in World History*. London: Pimlico, pp. 98–114.

Duncan, J.S. (2007) *In the Shadows of the Tropics: Climate, Race and Biopower in Nineteenth Century Ceylon*. Aldershot: Ashgate.

Economist, The (2010) 'Genesis Redux', *The Economist*, 20 May, www.economist.com/node/16163006.

Economist, The (2011) 'Welcome to the Anthropocene', *The Economist*, 26 May, www.economist.com/node/18744401.

Esposito, R. (2008) *Bios: Biopolitics and Philosophy*. Minneapolis: University of Minnesota Press.

Fitzsimmons, M. and Goodman, D. (1998) 'Incorporating nature: environmental narratives and the reproduction of food', in B. Braun and N. Castree (eds) *Remaking Reality: Nature at the Millennium*. London: Routledge, pp. 193–219.

Foster, J.B., Clark, B. and York, R. (2010) *The Ecological Rift: Capitalism's War on the Earth*. New York: Monthly Review Press.

Foucault, M. (1980) *The History of Sexuality. Vol. 1: An Introduction*. New York: Vintage Books.
Foucault, M. (2007) *Security, Territory, Population: Lectures at the Collège de France 1977–1978*. New York: Palgrave Macmillan.
Fukuyama, F. (2003) *Our Posthuman Future: Consequences of the Biotechnology Revolution*. New York: St. Martin's Press.
Gibson Graham, J.K. and Roelvink, G. (2009) 'An economic ethics for the Anthropocene', *Antipode*, 41(S1): 320–346
Giddens, A. (2009) *The Politics of Climate Change*. Cambridge: Polity.
Gregory, D. (2001) '(Post)colonialism and the production of nature', in N. Castree and B. Braun (eds) *Social Nature: Theory, Practice and Politics*. London: Blackwell, pp. 84–111.
Grove, R.H. (1995) *Green Imperialism: Colonial Expansion, Tropical Island Edens and the Origins of Environmentalism, 1600–1860*. Cambridge: Cambridge University Press.
Habermas, J. (2003) *The Future of Human Nature*. Cambridge: Polity.
Haraway, D. (1991) *Simians, Cyborgs and Women: The Reinvention of Nature*. London: Free Association Books.
Haslanger, S. (1995) 'Ontology and social construction', *Philosophical Topics*, 23(2): 95–125.
Hinchliffe, S. (2007) *Geographies of Nature*. London: Sage.
Jones, M. (2009) 'Phase space: geography, relational thinking, and beyond progress', *Progress in Human Geography*, 33(4): 487–506.
Kearns, G. (1998) 'The virtuous circle of facts and values in the New Western History', *Annals of the Association of American Geographers*, 88(3): 377–409.

Keen, D. (1994) *The Benefits of Famine: A Political Economy of Famine and Relief in Southwestern Sudan, 1993–1989*. Princeton, NJ: Princeton University Press.
Kiple, K. (2007) *A Movable Feast: Ten Millennia of Food Globalization*. Cambridge: Cambridge University Press.
Li, T.M. (2007) *The Will to Improve: Governmentality, Development, and the Practice of Politics*. Durham, NC: Duke University Press.
Liao, S.M., Sandberg, A. and Roache, R. (2012) 'Human engineering and climate change', *Ethics, Policy and the Environment*, 15(2): 206–221.
Mamdani, M. (1982) 'Karamoja: colonial roots of famine in north-east Uganda', *Review of African Political Economy*, 9(25): 66–73.
McKibben, B. (2006) *The End of Nature*. New York: Random House.
Mintz, S.W. (1986) *Sweetness and Power: the Place of Sugar in Modern History*. New York: Penguin Books.
Nally, D. (2011a) 'The biopolitics of food provisioning', *Transactions of the Institute of British Geographers*, 36(1): 37–53.
Nally, D. (2011b) *Human Encumbrances: Political Violence and the Great Irish Famine*. Notre Dame, IN: University of Notre Dame Press.
Nally, D. and Kearns, G. (2011) 'A closer look at famine: drought is only part of what's happening in East Africa', *Chronicle of Higher Education*, 58(9): 10–12.
Nixon. R. (2011) *Slow Violence and the Environmentalism of the Poor*. Cambridge, MA: Harvard University Press.
Parenti, C. (2011) *Tropic of Chaos: Climate Change and the New Geography of Violence*. New York: Nation Books.

Pelling, M. (2001) 'Natural disasters?' in N. Castree and B. Braun (eds) *Social Nature: Theory, Practice and Politics*. London: Blackwell, pp. 170–188.
Powell, J.M. (2000) 'Historical geographies of the environment', in B. Graham and C. Nash (eds) *Modern Historical Geographies*. London: Pearson, pp. 169–192.
Robbins, P. and Moore, S.A. (2012) 'Ecological anxiety disorder: diagnosing the politics of the Anthropocene', *Cultural Geographies*, 20(1): 3–19.
Rose, N. (2007) *The Politics of Life Itself: Biomedicine, Power, and Subjectivity in the Twentieth-First Century*. Princeton, NJ: Princeton University Press.
Said, E. (2000) *Reflections on Exile and Other Essays*. Cambridge, MA: Harvard University Press.
Schmidt, A. (1971) *The Concept of Nature in Marx*. London: New Left Review.
Shiva, V. (2010) 'Resources', in W. Sachs (ed.) *The Development Dictionary: A Guide to Knowledge as Power* (2nd edn). London: Zed Books, pp. 228–242.
Shiva, V. (2000) *Stolen Harvest: The Hijacking of the Global Food Supply*. Cambridge, MA: South End Press.
Simmons, I.G. (1998) '"To civility and to man's use": history, culture, and nature', *Geographical Review*, 88(1): 114–126.
Smith, N. (2006) 'There's no such thing as a natural disaster', 11 June, http://understandingkatrina.ssrc.org/Smith/.
Smith, N. (2008) *Uneven Development: Nature, Capital and the Production of Space*. Athens, GA: University of Georgia Press.
Sullivan, S. (2013) 'Banking nature? The spectacular financialisation of environmental conservation', *Antipode*, 45(1): 198–217.

Thompson, E.P. (1967) 'Time, work-discipline, and industrial capitalism', *Past and Present*, 38(1): 56–97.
Ullrich, O. (2010) 'Technology', in W. Sachs (ed.) *The Development Dictionary: A Guide to Knowledge as Power* (2nd edn). London: Zed Books, pp. 308–322.
Watts, M. (1983) *Silent Violence: Food, Famine, and Peasantry in Northern Nigeria* Berkeley: University of California Press.
Welzer, H. (2012) *Climate Wars: What People Will Be Killed For in the 21st Century*. Malden: Polity.
Whatmore, S. (2002) *Hybrid Geographies: Natures, Cultures, Spaces*. London: Sage.
Wood, E. M. (2000) 'The agrarian origins of capitalism', in F. Magdoff, J. B. Foster and F. H. Buttel (eds) *Hungry for Profit: the Agribusiness threat to Farmers, Food, and the Environment*. New York: Monthly Review Press, pp. 23–42.
Wright, R. (2005) *A Short History of Progress*. New York: Carroll & Graf Publishers.
Zinn, H. (1980) *A People's History of the United States, 1492–Present*. New York: Harper & Row.

FURTHER READING

Castree, N. and Braun B. (eds) (2001) *Social Nature: Theory, Practice, and Politics*. Oxford: Blackwell.
Chakrabarty, D. (2009) 'The climate of history: four theses', *Critical Inquiry*, 35: 197–222.

Dalby, S. (2007a) 'Ecology, security, and change in the Anthropocene', *Brown Journal of World Affairs*, 13(2): 155–164.
Smith, N. (2008) *Uneven Development: Nature, Capital and the Production of Space*. Athens, GA: University of Georgia Press.
Whatmore, S. (2002), *Hybrid Geographies: Natures, Cultures, Spaces*. London: Sage.

Section 8
The Production of Historical Geographical Knowledge

22 HISTORICAL GEOGRAPHICAL TRADITIONS

Ulf Strohmayer

Introduction

Of the many thematic (sub-)divisions customarily employed to render academic geography more manageable, 'historical geography' has had to contend more openly with the fact that its key goals entailed the amalgamation of traditions, methods and key practices from two rather traditional and well-established academic disciplines, namely geography and history. As such, attempts at writing historical geography inherited and combined cultural academic mindsets, as well as practices, that were routinely associated with particular intellectual endeavours. Concretely speaking, this initially amounted to an often static view of space being brought to bear on a dynamic notion of time; space, in other, more contemporary words, was temporalized geologically, while time was spatialized within a priori bounded and categorized borders. It is hence no surprise that the most 'natural' of convergence points between geography and history came initially, as Section 2 has demonstrated, in the form of the nation-state before attaching to other, commonly accepted geographical entities further 'down' the scales like regions or municipalities.

Academic Disciplines and Sub-disciplines

Of course, talk of 'well-established disciplines' in this broad context is to employ relative terms; it bears remembering that the majority of disciplines currently taught at school and university levels (and thus leaving an imprint onto a broader public imagination) have been established in the course of the 'long' nineteenth century, the exceptions possibly being theology, law and medicine. Prior to this drawn-out process of institutionally separating differentiated domains of knowledge many of the emerging and increasingly distinct categorical differences would have literally been 'un-thinkable'. In the present context, for instance, and commonly accepted though the distinction is by many today, the possibility of analytically prying away time from space would have seemed all but nonsensical to many knowledge-seeking individuals prior to the establishment of 'modern' university subjects (Wallerstein, 2001).

It is hence no surprise that for the longest time geography and geographers did not actually designate outwardly (and explicitly) any subdiscipline called 'historical geography' for the simple reason that allegedly autonomous realms of both space and time had not yet been conceptualized. In its stead, otherwise diverse notions such as Braudel's *genres de vie* or the German concept of *Kulturlandschaft* served those geographers with an interest in history; for the same reason it is difficult to find any self-designating 'historical geographers' until certainly well into the twentieth century when the idea of a 'geography of societal or socio-cultural change' (Dodgshon, 1998: 1) gradually acquired both a separate identity and currency within geographic discourses. When eventually it did emerge as such in the works of Paul Vidal de la Blache, Estyn Evans, Carl Sauer, H.C. Darby and others, their efforts – as discussed further on in this chapter – shared a methodological orientation towards meticulously descriptive and materially rooted forms of enquiry that came to set the norm for the majority of historical geographical scholarship ever since. It is in this form that 'historical geography' was taught especially at British and Irish universities from the 1930s onwards, with its practitioners traditionally expressing one of the more empirically minded, even 'scientific' of orientations, where the establishment of the former (in the form of 'evidence') is customarily thought to imply the latter status ('science'). The following two chapters will shed a more nuanced light on the multi-layered relationship between, on the one hand, the production of evidence and, on the other, ensuing geographical claims to knowledge; this chapter aims to script a framework for debates attaching to the production, justification and uses of geographical knowledge.

Towards a Historiography of Historical Geography

It is in this largely amorphous context that geography as a whole and historical geography in particular have come to develop their respective research agenda, themes and questions, many of which have been the subjects of the preceding chapters. In addition to these established, and thus in some way 'traditional' concerns, recent years have furthermore seen a proliferation of publications about the history of geography that can also loosely be attributed to 'historical geography'. Books or chapters in edited volumes documenting and analysing the development of the discipline (Livingstone, 1992; Gould and Strohmayer, 2004), its theories and ideas (Glacken, 1990; Peet, 1998; Martin, 2005), papers discussing the validity or the temporal location of 'history' or 'histories' of geography (Domosh, 1991; Driver, 1992; Buttimer, 1998; Mayhew, 2001) or essays offering explicit re-interpretations of key aspects or episodes in the 'history of geographic thought' (Wyly, 2009) have all added considerable scope and depth to historically motivated scholarship in geography, thus contributing greatly to an already well-established discourse attaching to the historiography of the discipline. While we lack the space properly to engage with this growing body of literature, it is well worth noting that such historiographies of geography tended to present a linear, progressing and thus necessarily streamlined storyline to their readers. Furthermore, since much of this work presupposes a 'safe (enough) distance' from that history for it to become an object of scholarly curiosity (Barnett, 1995), and the present volume clearly refuses to entertain debates about any such 'safety parameters', we gladly leave such contextual historiographies for others to write.

More important in the present perspective is the fact that any mention of context needs to be contextualized in turn; 'context', in other words, is not self-evident nor does it necessarily adopt a uniform scope and scale. The evident fact, for instance, that most traditions recognized in currently available historiographies of both geography and historical geography tend to be placed within the confines of specifically available linguistic boundaries and thus typically exclude lines of inquiry not at home in Western societies, may serve as a reminder that we should not take 'context' for granted. Even within the so-called 'Western' tradition, centrality is typically accorded to texts and networks of thought expressed in the English language, thus effectively flattening the landscape of possible intellectual encounters; this, we fear, may well become an even more dominant trend given the ongoing decline in linguistic competencies amongst British, American and Irish geographers. Furthermore, a similar 'flattening' of discursive possibilities is also happening outside the realm of English-speaking geographies by virtue of the academic prestige attaching to an individual's ability to import dominant trends from hegemonic discursive backgrounds into other linguistic traditions.

Landscape as 'Check-list'

That said, we can discern a number of communalities that have and – to varying degrees – continue to shape those key concepts and practices in historical geography that have been included in the present volume. Chief among these is a legacy that openly embraces the reconstruction of landscapes as the goal of knowledge construction in historical geography. Not dissimilar from archaeology in this respect, such 'reconstructive work' can take place in a diachronic, process-orientated, or synchronic, snapshot-like fashion (Bassin and Berdoulay, 2004). Broadly speaking, 'reconstruction' was the singular defining hallmark of work in historical geography up until the misleadingly so-called 'quantitative' revolution in the late 1950s and throughout the 1960s; since then, as we shall see soon, it has been challenged by a host of different approaches. However, prior to discussing any such epistemological concerns, one problem any 'reconstructive' form of historical geography ultimately faced was the sheer finitude of material to be mined in pursuit of knowledge. This problem is not unique to historical geography. In fact, it is even more pertinent in the sister sub-discipline of cultural geography, where similarly descriptive concerns and ideographic methods had led not merely to a 'check-list' mentality by the time the 1950s had arrived but had created an effectively 'over-grazed' empirical landscape. It is in this specific context that the scientific aspirations of historical geography became important: once a particular piece of historical geographical scholarship had been accepted, only the discovery and subsequent mining of new forms of evidence could alter a commonly accepted historical geography. In the absence of such novel data, research completed implied a 'mapped' form of historical geography that was no longer in need of scholarly attention.

If the above is perhaps a somewhat insensitive assessment, it nonetheless helps to explain the relative stasis of historical geography. Take, by way of example, one of the key manifestations of historically motivated scholarship emanating from geography: the study of urban morphologies. As practised especially by the late M.R.G. Conzen and enshrined into a concrete set of geographical practices in the so-called 'School' of Urban Morphology founded by his intellectual heir Jeremy Whitehand at the University of Birmingham, the effect of time onto space was often reduced to 'layerings' of material deposits with scant regard for the larger social, economic and political forces shaping the morphologies in question. Where these latter were accorded some prime of explanatory space, they were conceptualized as harmonizing with the material in question, rather than potentially conflicting with it. Furthermore, and congruent with a fundamental insistence on the importance of form and visually present evidence, this particular type of historio-geographical curiosity often took for granted the work of historians in establishing stable epochal categories – say 'Tudor England', 'Gilded Age US' or 'Weimar Germany' – and filled them with geographical specifics or variations. Crucially, such work was thought to be eminently transposable: the key elements – town plans, the form of both houses and urban settlements and land-use patterns – structurally transcended any spatial or temporal limits all the while evolving organically. The focus on cities in the Western orbit is no accident: Prior to the 1970s and 1980s, historical geography was largely a concern expressed by European geographers, with sojourns into non-European spaces few and far between.

Similar to a focus on urban morphologies is a widespread focus on the role and importance of regions and borders in the work of historical geographers. Here, too, what can be seen becomes an all too easy indicator or evidence for historical trends, influences and diffusion processes. Regions in particular have often assumed a role in geographical writings similar in degree of generalization accorded to the concept of 'period' in more narrowly construed historical writings (Wishart, 2004). All told, 'early' work in historical geography has thus bequeathed a materially rich tapestry of scholarship to future generations, which nonetheless increasingly parted company with key practices that had defined historical geography for a long time.

Historical Geography and the 'Spatial Turn'

It is difficult to pinpoint with any degree of exactitude what prompted historical geography to change. In marked contrast to other sub-disciplines within geography such as 'cultural geography' in particular, historical geography was less marked by dramatic shifts; rather, it appears to have drifted into oblivion only to be rescued by a sponge-like ability to absorb novel trends and ideas. If we had to identify a moment capable of summarizing, as well as influencing, change in the way geographers approached history, we would not hesitate to single out the 1984 publication of Denis Cosgrove's *Social Formation and Symbolic Change* as a defining event. Bringing a new vocabulary, as well as new methods and sources, to the study of historical geographies, the book insisted on the importance of a closer analysis of the means through which landscapes are constituted, disseminated and ultimately controlled. In turn, this motivated and legitimized scholarship on paintings or other forms of visual culture (Matless, 1996), photography (Ryan, 1994 and 2005; Schwartz, 1996) and maps (Edney, 1997) and thereby broadened not merely the scope of the material consciously used in historical geography but crucially embracing theoretically informed analyses of how different materialities, conditions of production and consumption impact upon the geographies and landscapes produced, re-produced and contested. The very fact that this latter term – 'contestation' – emerged as a valid category in work devoted to historical geography not only attests to a more theoretical (and thus less ideographic) orientation of such work but crucially points towards 'landscapes' that are progressively more heterogeneous and multi-layered. At the same time, the landscapes that continue to be the focus of much work in historical geography are increasingly contextualized within larger transformative processes like modernity (Ogborn, 1998), nationalism (Bassin, 1999), globalization or colonialism (Driver, 1992; Bonnett, 1997), just to mention some of the more recognized amongst the former – which not coincidentally all feature prominently in the present volume. The result has been a thorough and consciously sought-after de-naturalization of landscapes where previous scholarship often sought to emphasize those elements that made particular landscapes 'rhyme' or become congruent with, if not an outright expression of, a unity of historical purpose. Nowhere is this more apparent than in the context of nationalism: where previous generations of scholars had little difficulty identifying with and deploying terms such as say 'American', 'English', 'Irish' or 'German', often interpreting the geographies that rhyme with these terms as 'self-fulfilling' processes – in other words: reading the construction of landscapes in a purposeful manner, i.e. teleologically – present-day historical geographers acknowledge (qua 'contestation') that historical geographies could have developed differently (see Paasi, 1996 for a comprehensive example). One way of destabilizing the telos of traditional historical scholarship, as Simon Naylor has indirectly argued (2005), is to recognize the inherent mobility of many historical processes, is to recognize that alongside 'rooted' forms of materially present evidence relevant data can also circulate and in so doing can both influence a host of different geographical spaces and scales, all the while in turn being influenced by them. It is in the writings of Miles Ogborn (2007) and Charles Withers (2001), for example, that such increasingly nuanced geographies can emerge and fruitfully inform non-teleological scholarship in historical geography.

Humanism, Politics and Historical Geography

Abstracting from methodological considerations for the moment and recasting the above changes within a different context, we could place a more overt emphasis on the development, in recent years, of a more pronounced, albeit not undisputed, opening towards 'political' readings of past documents and developments. Associated to some degree with the incorporation of the work of Michel Foucault into the historical sciences – in the form both of a consciously adopted thematic context and as a mindset that embraces Foucault's notion of scholarship in the human sciences being primarily concerned with 'history of the present' (1977: 30) – 'history' is decreasingly conceptualized as an ongoing, progressing narrative. In its stead, geographers (and other scholars studying the human condition) adopted a view of history as a succession of differently structured orderings whose meaning is constituted through networks of epochally accepted statements. Geography, it was increasingly felt, played a crucial role in creating the conditions of possibility for the 'acceptance' of some statements as being 'true' all the while others did not achieve this distinction (Driver, 1985; Philo, 1992; Matless, 1992 and 1995). Given furthermore the undeniable historical richness of Foucault's empirical work, the scale of his reception in historical geography should surprise no one (Philo, 1987; Driver, 1989; Hannah, 1993).

In addition to this growing body of work it remains for us to point towards a no less important development in historical geography that is perhaps most overtly expressed in the works of Allan Pred (1990 and 1995) and Derek Gregory (1982). Both take their inspiration from a multiplicity of sources, including a humanist re-interpretation of the work of Karl Marx and the largely a-historical concept of 'time-geography' originally developed by Torsten Hägerstrand. The result focused consciously on the 'everyday' as a site for concrete historical struggles which thereby becomes formative for the construction of historically resonant forms of meaning. In this body of work, as well as that of others, historical geography emerges as a theoretically informed set of practices.

It is, however, well worth remembering at the same time that many geographers continue to use historical sources to support their arguments without necessarily accepting to be labelled 'historical geographers'. Looking over the work emanating from some of the more prolific and recognized of contemporary geographers – David Harvey, Neil Smith and Nigel Thrift come to mind – readers can observe the deployment of historical sources to quite varied ends. In a quite direct manner, their work, as well as the work of others, urges us to question, again and with renewed vigour, the validity and desirability, indeed the feasibility, of an earlier separation of 'space' from 'time' and to assert, following Dear's well-worn dictum, that 'the term "historical geography" is tautological, since all geography is (or should be) time- and place-specific' (1988: 270).

KEY POINTS

- During most of the time that Geography existed as a separate academic discipline, Historical Geography was not designated as a specific sub-discipline.
- In fact, many geographers would question the possibility of a meaningful separation of 'space' from 'time'.
- Most of the work designated by scholars as representing 'historical geography' concerned itself with the reconstruction of past landscapes.

- In recent years, and inspired by the 'cultural turn' characterizing most human sciences from the 1970s onwards, historically concerned geographical scholarship has analysed the means through which landscapes have historically been constructed.
- This work sought to de-naturalize dominant modes of constructing historical narratives by questioning commonly accepted links between time and space; teleological narratives that culminate in notions of 'progress', 'nation' or 'democracy' thus emerge as historically contingent and geographically bounded developments.

References

Barnett, C. (1995) 'Awakening the dead: who needs the history of geography?', *Transactions of the Institute of British Geographers*, 20: 417–419.

Bassin, M. (1999) *Imperial Visions: Nationalist Imagination and Geographical Expansion in the Russian Far East 1840–1865*. Cambridge: Cambridge University Press.

Bassin, M. and Berdoulay, V. (2004) 'Historical Geography: locating time in the spaces of modernity', in G. Benko and U. Strohmayer (eds) *Human Geography: A History for the 21st Century*. London: Arnold, pp. 64–82.

Bonnett, A. (1997) 'Geography, "race" and whiteness: invisible traditions and current challenges', *Area*, 29(3): 193–199.

Buttimer, A. (1998) 'Geography's contested stories: changing states-of-the-art', *Tijdschrift voor Economische en Sociale Geografie*, 89(1): 90–99.

Cosgrove, D. (1984) *Social Formation and Symbolic Change*. London: Croom Helm.

Dear, M. (1988) 'The postmodern challenge: reconstructing human geography', *Transactions of the Institute of British Geographers*, 13(3): 262–274.

Dodgshon, R. (1998) *Society in Time and Space: A Geographical Perspective on Change*. Cambridge: Cambridge University Press.

Domosh, M. (1991) 'Towards a feminist historiography of geography', *Transactions of the Institute of British Geographers*, 26(1): 95–104.

Driver, F. (1985) 'Power, space, and the body: a critical assessment of Foucault's Discipline and Punishment', *Environment and Planning D: Society and Space*, 3: 425–446.

Driver, F. (1989) 'The historical geography of the workhouse system in England and Wales, 1834–1883', *Journal of Historical Geography*, 15: 269–286.

Driver, F. (1992) 'Geography's empire: histories of geographical knowledge', *Environment and Planning D: Society and Space*, 10(1): 23–40.

Edney, M. (1997) *Mapping an Empire: The Geographical Construction of British India, 1765–1843*. Chicago: University of Chicago Press.

Foucault, M. (1977) *Discipline and Punish: The Birth of the Prison*. New York: Vintage.

Glacken, T. (1990) *Traces on the Rhodian Shore. Nature and Culture in Western Thought from Ancient Times to the End of the Eighteenth Century*. Berkeley, CA: University of California Press.

Gould, P. and Strohmayer, U. (2004) 'Geographical visions: the evolution of human geographic thought in the twentieth century', in G. Benko and U. Strohmayer (eds) *Human Geography: A History for the 21st Century*. London: Arnold, pp. 1–25.

Gregory, D. (I982) *Regional Transformation and Industrial Revolution: A Geography of the Yorkshire Woollen Industry*. London: Macmillan.

Hannah, M. (1993) 'Space and social control in the administration of the Oglala Lakota ("Sioux"), 1871–1879', *Journal of Historical Geography*, 19(4): 412–432.

Harvey, D. (1985) *Consciousness and the Urban Experience: Studies in the History and Theory of Capitalist Urbanisation*. Oxford: Blackwell.

Livingstone, D. (1992) *The Geographical Tradition*. Oxford: Blackwell.

Martin, G. (2005) *All Possible Worlds: A History of Geographical Ideas*. Oxford: Oxford University Press.

Matless, D. (1992) 'An occasion for geography: landscape, representation, and Foucault's corpus', *Environment and Planning D: Society and Space*, 10(1): 41–56.

Matless, D. (1995) 'Effects of history', *Transactions of the Institute of British Geographers*, 20(4): 405–409.

Matless, D. (1996) 'Visual culture and geographical citizenship: England in the 1940s', *Journal of Historical Geography*, 22: 424–439.

Mayhew, R. (2001) 'The effacement of early modern geography (c. 1600–1850)', *Progress in Human Geography*, 25(3): 383–401.

Naylor, S. (2005) 'Historical geography: knowledge, in place and on the move', *Progress in Human Geography*, 29: 626–633.

Ogborn, M. (1998) *Spaces of Modernity: London's Geographies 1680–1780*. London: Guilford.

Ogborn, M. (2007) *Indian Ink: Script and Print in the Making of the English East India Company*. Chicago: University of Chicago Press.

Paasi, A. (1996) *Territories, Boundaries and Consciousness: The Changing Geographies of the Finnish–Russian Border*. Chichester: Wiley.

Peet, R. (1998) *Modern Geographic Thought*. Oxford: Blackwell.

Philo, C. (1987) '"Fit localities for an asylum": the historical geography of the nineteenth-century "mad business" in England as viewed through the pages of the Asylum Journal', *Journal of Historical Geography*, 13: 398–415.

Philo, C. (1992) 'Foucault's geography', *Environment and Planning D: Society and Space* 10(2): 137–161.

Pred, A. (1990) *Lost Words and Lost Worlds: Modernity and the Language of Everyday Life in Late Nineteenth-Century Stockholm*. Cambridge: Cambridge University Press.

Pred, A. (1995) *Recognizing European Modernities: A Montage of the Present*. London: Routledge.

Ryan, J. (1994) 'Visualizing imperial geography: Halford Mackinder and the Colonial Office Visual Instruction Committee, 1902–1911', *Ecumene*, 1: 157–176.

Ryan, J. (2005) 'Photography, visual revolutions, and Victorian Geography', in D.N. Livingstone and C. Withers (eds) *Geography and Revolution*. Chicago: University of Chicago Press, pp. 199–238.

Schwartz, J. (1996) 'The geography lesson: photographs and the construction of imaginative geographies', *Journal of Historical Geography*, 22: 16–45
Wallerstein, I. (2001) *Unthinking Social Science: The Limits of 19th Century Paradigms*. Philadelphia, PA: Temple University Press.
Wishart, D. (2004) 'Period and region', *Progress in Human Geography*, 28(3): 305–319.
Withers, C. (2001) *Geography, Science and National Identity: Scotland since 1520*. Cambridge: Cambridge University Press.
Wyly, E. (2009) 'Strategic positivism', *The Professional Geographer*, 61(3): 310–322.

FURTHER READING

Barnett, C. (1995) 'Awakening the dead: who needs the history of geography?', *Transactions of the Institute of British Geographers*, 20: 417–419.
Bassin, M. and Berdoulay, V. (2004) 'Historical Geography: locating time in the spaces of modernity', in G. Benko and U. Strohmayer (eds) *Human Geography: A History for the 21st Century*. London: Arnold, pp. 64–82.
Domosh, M. (1991) 'Towards a feminist historiography of geography', *Transactions of the Institute of British Geographers*, 26(1): 95–104.
Gould, P. and Strohmayer, U. (2004) 'Geographical visions: the evolution of human geographic thought in the twentieth century', in G. Benko and U. Strohmayer (eds) *Human Geography. A History for the 21st Century*. London: Arnold, pp. 1–25.
Livingstone, D. (1993) *The Geographical Tradition*. Oxford: Blackwell.

23 ILLUSTRATIVE GEOGRAPHIES

John Morrissey

Introduction

Researching and writing any historical geography involves envisioning and (re)presenting in the present a particular world from the past. In accessing that past, we use various forms of evidence to narrate the historical geography at the heart of our inquiry. Frequently, this also involves an attempt to illustrate that geography by using a contemporary map, sketch, image or other visual signifier. Visualization is after all 'at the heart of geographic practice' (Aitken and Craine, 2005: 251). However, it is not just in the *depiction* of the past that visual representations can be used; they can also be fruitfully utilized as *sources* in their own right to deconstruct the same diverse geographies that are typically divulged from analysis of conventional documentary archives. This chapter explores the use of visual sources in the production of historical geographical research. By thinking broadly about *illustrative geographies*, the aim is to signal the diverse visual mediums through which historical geographies can be envisioned and represented before reflecting on appropriate visual analytical techniques.

Maps and Map-making in the Cartographic Tradition

One of the most important sources of visual representation is the map. As Chris Perkins (2003: 344) notes, the 'ability to construct and read maps is one of the most important means of human communication, as old as the invention of language'. The emergence of geography as a university discipline in the late nineteenth century coincided with cartography playing a central role in the envisioning and mapping of western colonialism. In the age of high imperialism, leading geographers such as Halford Mackinder in England and Friedrich Ratzel in Germany were active players in the project of imperial science that underpinned and supported colonial expansion (Livingstone, 1992; Heffernan, 2000; Driver, 2001). The geographical tool of the map 'provided the European imperial project with arguably its most potent device' (Heffernan, 2003: 11; see also Chapter 1). It is important, therefore, not to see cartography as historically merely a discursive method of knowledge accumulation and display but rather as a key practice that facilitated what Edward Said terms 'acts of geographical violence', through which spaces were 'explored, charted, and finally brought under control' (Said, 1993: 271). Contextualizing the original use of maps can tell us much about the society, institutions and ideologies that produced them. In other words, maps are 'not just artifacts' because 'mapping is a *process* reflecting a way of thinking' (Perkins, 2003: 343).

Traditionally, the *positivist* tradition of making and reading maps reflected a Cartesian belief in the human ability to objectively mimic the *real* world in its *representation*. The work of Brian Harley in the 1980s, however, brought about a sea change in the discipline of cartography, with his critique of positivism bringing maps into much broader debates in the arts, humanities and social sciences concerning discourse, power and knowledge (Harley, 1988, 1989a, 1989b). Drawing on the writings of Erwin Panofsky, Michel Foucault and Jacques Derrida, Harley divulged cartographic representations as socially constructed *images* with historically specific *codes*. Grounded in semiotics, his work outlined the shortcomings of the perception that cartography can 'mirror accurately some aspect of "reality" which is simple and knowable and can be expressed as a system of facts' (Harley, 1989b: 82). Instead, he demonstrated the various modalities of power and authority that function cartographically to script particular social relations of imperialism, race, gender and so on (see also: Wood, 1993; Dorling and Fairbairn, 1997; Perkins, 2003). For Harley, the father of *critical cartography*, maps are ultimately a 'teleological discourse, reifying power, reinforcing the *status quo*, and freezing social interaction within chartered lines', and any analysis that ignores the political and cultural connotations of cartographic representation is in effect 'ahistorical' (1988: 302–303).

Harley's focus on the deconstruction of the political and cultural symbolism of cartography has been instrumental in the broader problematization of all forms of geographical representation. Aitken and Craine (2005: 254) recently observe, for example, that the 'interpretation of geographical representations provides insights into how society and space are ordered and how the construction and representation of that space is manipulated by powerful groups through cultural codes that promote dominant ideologies' (Aitken and Craine, 2005: 254). Others whose work in critical cartography has had a significant impact in historical geography include John Pickles (1992) and Denis Wood (1993). Pickles has made excellent use of hermeneutics to offer a textual analysis of propaganda maps, while Wood's exploration of the communicative power of maps has emphasized the 'importance of understanding the signs and myths which are embodied in cartography codes' (Aitken and Craine, 2005: 240).

In more recent years, *historical geographic information systems (GIS)* have been used in the study of historical geography in a variety of contexts (see, for example, the special volume 33 of *Historical Geography*, which includes work by Wendy Bigler, Don DeBats and Mark Lethbridge, and Paul Ell and Ian Gregory). In historical GIS, source data is typically converted from archival analog form to digital format, and, as Anne Kelly Knowles (2000: 452) points out, the accumulation of digital spatial databases is creating various opportunities for 'GIS analyses of the recent past, particularly in urban, transportation, business, and environmental history' (see also Ell and Gregory, 2008). Knowles's plea for the importance of historical GIS in historical geographical research centres on the contention that '[m]apping data reveals dimensions of historical reality and change that no other mode of analysis can reveal' (2000: 453). Several anxieties, however (concerning the same issues of authority, power and knowledge that need to be considered when using conventional cartography – as outlined above), have been expressed about the unproblematic use of GIS in historical geographical research (see, for example: Gregory, 1994; Curry, 1995; and Pickles, 1995).

Visualizing Landscape

One of the key areas of research in historical geography is the study of past cultural landscapes. The remnants of many historical built environments are often still present in the contemporary landscape, but overlays of newer structures typically limit the extent to which we can access and interrogate them. In formulating a picture of a specific space in a particular time in the past, contemporary representations – written or visual – provide key lenses through which we can envision. And in envisioning landscape, a variety of *visual sources*, as outlined by Iain Black (2003: 487), are all valid archives to be fruitfully engaged: 'Paintings, sketches, engravings and architectural drawings are all valuable in the essentially archaeological practice of recovering the symbolic geography of past landscapes.'

Over the last two decades or more, landscape studies in historical geography have seen a move away from the positivist and largely descriptive nature of the Sauerian school of landscape interpretation (see Chapter 13) to more contextualized and theoretically situated forms of interrogation, where the landscape is seen as a *text* to be *read*. As Aitken (2005: 241) argues, rather than simply seeing landscape as a given, geographers now acknowledge that 'places are actively produced and struggled over', and it is the interrogation and deconstruction of that process that holds the key to understanding any given landscape and its multiple layers and complexities. To this end, new forms of *textual analysis*, drawing upon *literary metaphor* and *iconography*, for example, have come to the fore (Cosgrove and Daniels, 1988; Barnes and Duncan, 1992; Cosgrove and Domosh, 1993). The late Denis Cosgrove (1993), for example, has used the work of Italian architect Andrea Palladio to interrogate the visual production of cultural meaning through landscape. In his examination of the *envisioning* of landscape in sixteenth-century northern Italy, he asserts that 'to understand how human groups come to terms with and transform their material environments [it is necessary] to pay as much attention to the intellectual forces and spiritual sensibilities that empower those groups as to the economic, social and environmental constraints with which they subsist' (1993: xiii).

Using Images in Historical Geography

The use of visual sources is not just confined to historical work on landscape; images are a rich mine of knowledge in the study of historical geography more generally. Key sources range from historical photographs, posters and postcards to diagrams, sketches and woodcuts. They also include tele-visual sources such as cartoons and film, for instance, and it is important to recognize all of these as archives requiring the same care and systematic assessment as is expected when using more conventional written sources. As Gillian Rose (2007: 262) argues, '[p]recisely because images matter, because they are powerful and seductive, it is necessary to consider them critically' (Rose, 2007: 262). In our image-saturated world, no visual imagery is ever *innocent*, and therefore contextualization is of paramount importance (Rogoff, 2000; Schwartz and Ryan, 2003; Mirzoeff, 2005). Analysing images, furthermore, must not concern itself with merely *how images look* but in addition *how they are looked at.*

Visual images *affect* us in myriad ways and play key roles in the discursive networking of power/knowledge, and the envisioning of social categories and hierarchies (Hall, 1997). As Gordon Fyfe and John Law (1988: 1) argue:

> To understand a visualisation is [...] to enquire into its provenance and into the social work that it does. It is to note its principles of inclusion and exclusion, to detect the roles that it makes available, to understand the way in which they are distributed, and to decode the hierarchies and differences that it naturalizes.

Since the 1990s, various historical and cultural geographers have highlighted the neglect of the study of the power relations embedded in visual archives – such as advertising, art works, photography and movies (Cresswell and Dixon, 2002; Aitken and Craine, 2005). In more recent years, critical attention has been paid to the visual 'spaces of display and performance' of, for example, art galleries, museums and the theatre (Blunt, 2003: 85). Yet, despite this, as Cresswell and Dixon (2002) note in their geographical engagement with film, we do not possess the same critical faculties when reading vision as we do when reading the written word (for valuable work outside of geography on film and representation, see: Jowett and Linton, 1989; Carnes, 1996). Rob Bartram (2003: 150) outlines the resulting key challenge:

> while visual imagery has formed a dominant way of expressing geographical knowledges, our interpretation of visual imagery has sometimes lacked a critical awareness: all too often, visual imagery has been used as a straightforward reflection of reality, with no sense of how, when and by whom the image has been produced.

Bartram goes on to highlight how the interpretation of visual imagery is 'intricately linked to philosophical and theoretical debates about the production and experience of culture' and echoes calls made elsewhere for a systematic and critical engagement with all forms of visual sources (2003: 158; see also: Hannam, 2002; Aitken and Crane, 2005; Rose, 2007).

Visual Methodologies

So what techniques are required for using visual images? Aitken and Craine (2005) employ four types of visual images (art, photography, film and advertising) to offer four different methods of analysis: semiotic, feminist, psychoanalytic and discourse analysis. Rob Bartram (2003) also presents a succinct and cogent explication of how to carry out a systematic examination of the relationship between *signs* and *signification* at the heart of visual imagery (for useful work outside of geography on representation and visual cultures, see: Lacey, 1998; Alvarado et al., 2000; Thwaites et al., 2002; and Croteau and Hoynes, 2003). Gillian Rose's work on visual imagery, however, is perhaps the most important reference point for historical geographers (Rose, 1996, 2007). In her 2007 comprehensive exploration of *visual methodologies* in geography, she signals the choice of techniques available in the analysis of visual materials: *compositional interpretation*; *content analysis*; *semiology* (or *semiotics*); *psychoanalysis*; and *discourse analysis* (which she outlines as two distinct forms of inquiry – one focused on text and intertextuality, with the other concerned with institutions and ways of seeing). Whatever analytical technique is chosen, she urges the development of a critical visual methodology; one which requires an acknowledgment of 'the differentiated effects of both an image's way of seeing and [our] own' (Rose, 2007: 262).

Rose identifies three *sites* at which the meanings of images are made and effected: 'the site of *production*, the site of the *image or object itself* and the site of *audiencing*' (Rose, 2007: 257). For all three sites, Rose highlights three *modalities* that can be critically read to enable a deeper understanding of any given image: the *technological* modality, the *compositional* modality and the *social* modality:

> The technological concerns the tools and equipment used to make, structure and display an image; the compositional concerns the visual construction, qualities and reception of an image; and the social concerns the social, economic, political and institutional practices and relations that produce, saturate and interpret an image. (2007: 258)

Although the distinctions drawn above are more fluid in practice (Rose acknowledges that there are many overlaps between sites and modalities), the working paradigm affords a valuable way in which to approach the use of images in historical geographical research.

In practical terms, the useful distinction of three *sites* through which an image's meanings are made allows for specific aspects of an image's (i) construction, (ii) medium and (iii) consumption to be honed in on in accordance with particular research concerns:

(i) In concentrating on the *construction* of the image – how meaning is encoded though the process of production – key basic questions include (see Rose, 2007: 258–259 for a fuller set of starting questions for visual research):

- When was the image made?
- Where was the image made?
- How was the image made?

(ii) When concerned with the image or *medium* itself – its components and characteristics – practical questions we could ask include:

- What are the image's components and how are they arranged?
- What is the vantage point of the image and how is the viewer's eye directed?
- What is the genre of the image and what are its key elements?

(iii) Finally, if the focus of research is the *consumption* of the image – how audiences decode meaning in various ways – useful questions include:

- Who were the original audience(s) for the image?
- How do different audiences interpret the image?
- What were the image's geographies of display and re-display over time?

The specific questions we wish to ask of our evidence may also of course be informed by broader theoretical concerns. If we were using elements of postcolonial theory, for example, our questions might involve teasing out configurations of colonial social binaries. If our archival work was also critically informed by feminist theory, attention might additionally be directed to issues of gendered social hierarchies. Whatever our theoretical framework, however, a key overarching methodological concern in critically analysing visual archives should be to ask questions of the evidence that seek to historicize, spatialize and contextualize meaning.

Conclusion

This chapter has sought to outline some of the key issues in using visual sources in historical geography. Visual archives are rich sources for historical geographers. They provide exciting ways in which to envision and interrogate the human geographies of the past. There are dangers of relying too heavily on visual sources. Hannam (2002: 190), for example, points to important Marxist and feminist critiques of abstracting research beyond the realm of material and corporeal conditions when privileging the representational and the symbolic (see also Smith, 1993 and Aitken, 2005). Addressing this issue, Iain Black cites the work of Derek Holdsworth to argue for 'the need to combine analyses of visual evidence with a wider set of documentary sources to develop a fuller understanding of the creation and functioning of built environments in the past' (Holdsworth, 1997; Black, 2003: 489). Indeed, many historical geographers use visual data in conjunction with other forms of written, oral and material sources, especially given that all research contexts are ultimately dependent on limited available evidence; a point taken up in the concluding chapter. The key concern, however, in using visual sources in historical geography is to be systematic and critical in contextualization, interrogation and representation.

KEY POINTS

- It is not just in our *depiction* of the past that visual representations can be used; they can also be fruitfully utilized as *sources* in their own right.
- In critical cartography, the work of Brian Harley has demonstrated the various modalities of power and authority that function to script particular social relations.

- In more recent years, *historical GIS* has been used in the study of historical geography in a variety of contexts.
- Images are a rich mine of knowledge in the study of historical geography; in envisioning landscape, for example, a variety of *visual sources* such as paintings, sketches and architectural drawings have been fruitfully engaged.
- In the analysis of images, there are a variety of techniques available such as compositional interpretation; content analysis; feminist analysis; psychoanalysis; semiotics; and discourse analysis.
- Gillian Rose's (2007) useful distinction of three *sites* through which an image's meanings are made allows for specific aspects of an image's (i) construction, (ii) medium and (iii) consumption to be explored.
- A key overarching methodological concern in critically analysing visual images in historical geography should be to ask questions of the evidence that seek to historicize, spatialize and contextualize meaning.

References

Aitken, S. (2005) 'Textual analysis: reading culture and context', in R. Flowerdew and D. Martin (eds) *Methods in Human Geography* (2nd edn). Harlow: Pearson, pp. 233–249.

Aitken, S. and Craine, J. (2005) 'Visual methodologies: what you see is not always what you get', in R. Flowerdew and D. Martin (eds) *Methods in Human Geography* (2nd edn). Harlow: Pearson, pp. 250–269.

Alvarado, M., Buscombe, E. and Collins, R. (eds) (2000) *Representation and Photography: A Screen Education Reader*. New York: Palgrave.

Barnes, T. and Duncan, J. (eds) (1992) *Writing Worlds: Discourse, Text and Metaphor in the Representation of Landscape*. London: Routledge.

Bartram, R. (2003) 'Geography and the interpretation of visual imagery', in N. Clifford and G. Valentine (eds) *Key Methods in Geography*. London: Sage, pp. 149–159.

Bigler, W. (2005) 'Using GIS to investigate fine-scale spatial patterns in historical American Indian agriculture', *Historical Geography*, 33: 14–32.

Black, I.S. (2003) 'Analysing historical and archive sources', in N.J. Clifford and G. Valentine (eds) *Key Methods in Geography*. London: Sage, pp. 475–500.

Blunt, A. (2003) 'Geography and the humanities tradition', in S.L. Holloway, S.P. Price and G. Valentine (eds) *Key Concepts in Geography*. London: Sage, pp. 73–91.

Carnes, M. (ed.) (1996) *Past Imperfect: History According to the Movies*. London: Cassell.

Cosgrove, D. (1993) *The Palladian Landscape: Geographical Change and its Cultural Representations in 16th Century Italy*. University Park, PA: Pennsylvania State University Press.

Cosgrove, D. and Daniels, S. (eds) (1988) *The Iconography of Landscape: Essays on the Symbolic Representation, Design and Use of Past Environments*. Cambridge: Cambridge University Press.

Cosgrove, D. and Domosh, M. (1993) 'Author and authority: writing and the new cultural geography', in J. Duncan and D. Ley (eds) *Place/Culture/Representation*. London: Routledge, pp. 25–38.

Cresswell, T. and Dixon, D. (2002) 'Introduction: engaging film', in T. Cresswell and D. Dixon (eds) *Engaging Film: Geographies of Mobility and Identity*. Oxford: Rowman & Littlefield, pp. 1–10.

Croteau, D. and Hoynes, W. (2003) *Media Society: Industries, Images and Audiences*. London: Sage.

Curry, M. (1995) 'Rethinking the rights and responsibilities of geographic information systems: beyond the power of the image', *Cartography and Geographic Information Systems*, 22(1): 58–69.

DeBats, D.A. and Lethbridge, M. (2005) 'GIS and the American city: nineteenth century residential patterns', *Historical Geography*, 33: 78–98.

Dorling, D. and Fairbairn, D. (1997) *Mapping: Ways of Representing the World*. Harlow: Longman.

Driver, F. (2001) *Geography Militant: Cultures of Exploration and Empire*. Oxford: Blackwell.

Ell, P.S. and Gregory, I.N. (2005) 'Demography, depopulation and devastation: exploring the geography of the Irish Potato Famine', *Historical Geography*, 33: 54–75.

Ell, P.S. and Gregory, I.N. (2008) *Historical GIS: Technologies, Methodologies, and Scholarship*. Cambridge: Cambridge University Press.

Fyfe, G. and Law, J. (1988) *Picturing Power: Visual Depictions and Social Relations*. London: Routledge.

Gregory, D. (1994) *Geographical Imaginations*. Oxford: Blackwell.

Hall, S. (1997) 'The work of representation', in S. Hall (ed.) *Representation: Cultural Representations and Signifying Practices*. London: Sage, pp. 13–74.

Hannam, K. (2002) 'Coping with archival and textual data', in P. Shurmer-Smith (ed.) *Doing Cultural Geography*. London: Sage, pp. 189–197.

Harley, J.B. (1988) 'Maps, knowledge and power', in D. Cosgrove and S. Daniels (eds) *The Iconography of Landscape: Essays on the Symbolic Representation, Design and Use of Past Environments*. Cambridge: Cambridge University Press, pp. 277–312.

Harley, J.B. (1989a) 'Deconstructing the map', *Cartographica*, 26(2): 1–20.

Harley, J.B. (1989b) 'Historical geography and the cartographic illusion', *Journal of Historical Geography*, 15(1): 80–91.

Heffernan, M. (2000) 'Fin de siècle, fin du monde: on the origins of European geopolitics, 1890–1920', in K. Dodds and D. Atkinson (eds) *Geopolitical Traditions: A Century of Geopolitical Thought*. London: Routledge, pp. 27–51.

Heffernan, M. (2003) 'Histories of geography', in S.L. Holloway, S.P. Price and G. Valentine (eds) *Key Concepts in Geography*. London: Sage, pp. 3–22.

Holdsworth, D.W. (1997) 'Landscape and archives as texts', in P. Groth and T. Bressi (eds) *Understanding Ordinary Landscapes*. New Haven, CT: Yale University Press, pp. 44–55.

Jowett, G. and Linton, J.M. (1989) *Movies as Mass Communication*. London: Sage.
Knowles, A.K. (2000) 'Introduction: historical GIS: the spatial turn in social science history', *Social Science History*, 24(3): 451–470.
Lacey, N. (1998) *Image and Representation: Key Concepts in Media Studies*. New York: Palgrave.
Livingstone, D. (1992) *The Geographical Tradition: Episodes in the History of a Contested Enterprise*. Oxford: Blackwell.
Mirzoeff, N. (2005) *Watching Babylon: The War in Iraq and Global Visual Culture*. New York: Routledge.
Perkins, C. (2003) 'Cartography and graphicacy', in N.J. Clifford and G. Valentine (eds) *Key Methods in Geography*. London: Sage, pp. 343–368.
Pickles, J. (1992) 'Text, hermeneutics and propaganda maps', in T. Barnes and J. Duncan (eds) *Writing Worlds: Discourse, Text and Metaphor*. London: Routledge, pp. 193–230.
Pickles, J. (ed.) (1995) *Ground Truth: The Social Implications of Geographic Information Systems*. New York: Guilford Press.
Rogoff, I. (2000) *Terra Inferma: Geography's Visual Culture*. London: Routledge.
Rose, G. (1996) 'Teaching visualised geographies: towards a methodology for the interpretation of visual materials', *Journal of Geography in Higher Education*, 20(3): 281–194.
Rose, G. (2007) *Visual Methodologies: An Introduction to the Interpretation of Visual Materials* (2nd edn). London: Sage.
Said, E. (1993) *Culture and Imperialism*. New York: Alfred A. Knopf.

Schwartz, J. and Ryan, J. (eds) (2003) *Picturing Place: Photography and the Geographical Imagination*. London: I.B.Tauris.
Smith, N. (1993) 'Homeless/global: scaling places', in J. Bird, B. Curtis, T. Putman, G. Robertson and L. Tickner (eds) *Mapping the Futures: Local Cultures, Global Change*. London: Routledge, pp. 87–119.
Thwaites, T., Davis, L. and Mules, W. (2002) *Introducing Cultural and Media Studies: A Semiotic Approach*. New York: Palgrave.
Wood, D. (1993) *The Power of Maps*. London: Routledge.

FURTHER READING

Harley, J.B. (1988) 'Maps, knowledge and power', in D. Cosgrove and S. Daniels (eds) *The Iconography of Landscape: Essays on the Symbolic Representation, Design and Use of Past Environments*. Cambridge: Cambridge University Press, pp. 277–312.
Pickles, J. (1992) 'Text, hermeneutics and propaganda maps', in T. Barnes and J. Duncan (eds) *Writing Worlds: Discourse, Text and Metaphor*. London: Routledge, pp. 193–230.
Rose, G. (2007) *Visual Methodologies: An Introduction to the Interpretation of Visual Materials* (2nd edn). London: Sage.
Schwartz, J. and Ryan, J. (eds) (2003) *Picturing Place: Photography and the Geographical Imagination*. London: I.B.Tauris.

24 EVIDENCE AND REPRESENTATION

John Morrissey

Introduction

This final chapter focuses on the key determinant of doing any research: the *evidence*. Research in human geography involves the employment of evidence from any number of sources in the contemporary world; all of which pose significant challenges. Historical geographical research, however, incorporates yet more challenges given that its (re)presentation relies on the remnants, records and recollections of past worlds that we can no longer personally encounter or verify. From Alan Baker's plain observation that 'the dead don't answer questionnaires', we can begin to reflect on the difficulties of evidence in researching and writing historical geography (Baker, 1997). This chapter outlines some of the key challenges of doing historical geography by concentrating on four crucial aspects of research and representation: the field, sources, interpretation and narration.

The Field

Traditionally, historical geography has been concerned with a number of important issues in its *field* of study: from population and migration to colonial expansion and settlement; from agriculture and agrarian change to industrialization and urban morphology. Within this field, a 'long-standing empirical tradition' of doing research has been focused on 'reconstructing' past environments and highlighting material historical changes in landscape and society (Nash and

Graham, 2000: 3). However, recent years have witnessed new perspectives and themes within historical geography that have been consonant with a broader concern in human geography for theorizing the production of knowledge about nature and culture. This has resulted in three key developments: (i) a widening of what constitutes the historical geographical field; (ii) recognition of the diversity of meaning within it; and (iii) subsequently a closer theorization of context and situatedness in the narration of historical geographical knowledge.

Thematically, historical geography has seen a recent broadening of research foci, with new political, cultural and symbolic questions of power, ideology, race and gender, for example, being developed from traditional concerns with the material worlds of the past. This has been achieved by theorizations of historical spaces that recognize the complexities of contested pasts, which in turn require careful situated interrogation. The notion of *situated knowledges* of the geographical field (which had come to the fore in human geography by the mid-1990s) resulted in work in historical geography becoming more differentiated, contextualized and self-reflexive of the theoretical and methodological production of knowledge (Ogborn, 1996). This development brought into sharp relief the limitations of atheoretical historical geography or what has been described as the *self-evident* school (Harley, 1989). In historical landscape studies, for example, George Henderson (1998: 94–95) notes that the move to situated knowledges about nature and culture was 'rather grim news for the "self-evident" school', arguing instead that 'the more the meaning of landscape moves away from object toward context the more powerful it is as a concept for the study of social life'.

Efforts to situate more nuanced knowledges of the past in historical geography has coincided with a growing reflection on *fieldwork* and how its *practice* has generated specific historical geographical knowledges over time. Hayden Lorimer (2003: 197), for example, has shown how 'the practice of learning geography, and the arenas in which knowledge-making takes place, can be usefully positioned within changing histories of the discipline'. Elsewhere, with Nick Spedding, he has used the example of one family's expedition and field trip to Glen Roy, Scotland, in the 1950s to highlight the import of the embodied practices of fieldwork, such as 'travelling' and 'residing', which serve ultimately to alter 'the ways in which the site of scientific investigation is experienced and understood' (2005: 13).

Sources

What comprises the field in historical geography has of course implications for how that field is encountered in terms of evidence. Recent debates in geography more generally about 'what constitutes the "field" and the practice of "fieldwork"' has occurred in conjunction with a 'growing respect for methods that are less constrained by "field-orientated" empiricism and the "rigour" of science' (Aitken, 2005: 235, see also Keighren, 2012). In historical geography, the development of a wider thematic field has coincided with the emergence of a broader delineation of what constitutes valid *sources*; whose diversity has in turn prompted the employment of new techniques for their effective examination (including more qualitative methods such as textual analysis, discourse analysis and semiotic analysis, as discussed in the previous chapter; see also Hannam, 2002b; Shurmer-Smith, 2002; Black, 2003).

The study of historical geography can be based on a wide range of written, visual, oral and material sources. As Miles Ogborn (2003: 101) notes, there is a 'huge variety of historical data', spanning various forms of 'written and numerical material', 'oral history', 'dendrohistory', 'visual images' and 'maps'. Kevin Hannam (2002a) distinguishes between *formal* and *informal* sources (others make the distinction of *official* and *unofficial* sources; see Cloke et al., 2004). For Hannam, formal sources encompass national archive materials, census records, state documents and other official governmental and institutional documentation in the public realm. Informal sources include *published* materials such as memoirs, chronicles and biographies, and *unpublished* archives not produced for public consumption such as letters, diaries and photographs. Hannam's working distinction is useful if only to remind us of the alternatives available when attempting to recover the human geographies of the past that do not feature prominently (or are perhaps invisible) in formal historical records. Indeed the limited scope of employing solely official sources has prompted historical geographers to increasingly use what Cloke et al. (2004: 93) term *imaginative sources*. The imaginative use of such sources as literature, travel writing, newspapers, cinema, photography and electronic media are now commonplace in the sub-discipline.

Using any source material in historical geography places particular methodological demands on the researcher. It is vital to critically evaluate sources prior to employing them. Iain Black (2003: 479–480) identifies four key considerations to bear in mind: (i) the authenticity of the source; (ii) the accuracy of the source; (iii) the original purpose of the data; and (iv) how the specific archiving imposes a particular classification and order upon historical events. Thinking through questions of accuracy, authenticity, political and ideological underpinnings and power relations (which are often implicit) for all sources is an essential part of doing historical geography. These issues, together with the persistent problem of incomplete, sparse or non-existent sources, render *cross-referencing* wherever possible an invaluable component of effective research. Kevin Hannam (2002b: 191), for example, reminds us that since all sources are 'to some degree inaccurate, incomplete, distorted or tainted', the best way to 'understand the significance of an event or a representation' is to appraise 'as wide a range of sources as possible'. Finally, as outlined in the previous chapter on visual sources, the effective examination of any source depends upon a careful and informed interpretation that attempts to historicize and spatialize context and meaning.

Interpretation

Historical geographers have long reflected on the unavoidable problem of partiality or bias in source materials. Jim Duncan (1990), for example, in his exploration of the Kandyan Kingdom in nineteenth-century Sri Lanka, observed the interpretive problems of his sources, outlining issues of officialdom, power relations, empowerment and dispossession. Others, too, have underlined the historiographic asymmetries between dominant elite voices and marginalized powerless ones in foremost, official government archives (Withers, 2002; Ogborn, 2003; Morrissey, 2004). Ignoring this issue, as Kevin Hannam (2002a: 115) points out, typically results in the following scenario: 'In uncritically adopting an "official" government archive as the primary source of knowledge, a researcher may adopt the view of the established government and ignore, or at the very least treat as secondary, the voices of marginalized people.'

In the colonial context, the difficulties of examining the political, social and cultural productions of those opposed to a dominant hegemony (productions of the colonized as opposed to productions imposed on them) has been identified by various historical geographers as a central research concern (Wishart, 1997; Nash and Graham, 2000; Black, 2003). In many cases, the absence of *official* historical documentation of particular forms of resistance, for instance, does not necessarily mean that reactions were undemonstrative. However, it is important to bear in mind Jim Duncan's caveat on this point: that 'overly ambitious theorizing' can serve to 'obscure and bury even deeper indigenous knowledges and practices' (1999: 127). Cross-referencing other kinds of sources (such as diary accounts, poetry and oral history, for example) is perhaps a more effective way of illuminating skewed pictures that emerged from official empirical archives.

A prominent critique of traditional historical geography is that its dominant empirical tradition of doing research has been largely atheoretical (see Chapter 22). With little theorization of key concepts such as landscape and culture, for example, a self-evident nature of explanation prevailed, which often served to negate the intricate complexities of how places and events were experienced differently through history – depending on a multitude of issues such as class, race, gender, disability and so on. However, it is neither entirely fair nor fruitful to posit traditional historical geography as atheoretical. The sub-discipline has a long-established tradition that reflects theoretical influences from cognate disciplines and more broadly social theory. These have altered and continued through time and indeed, as Kevin Hannam (2002a: 113) notes, '[h]istorical cultural geography based on archival research has arguably been at the forefront of [recent] theoretical advances in geography'.

Broader theoretical concerns always form an integral part of interpreting sources. Even those that claim a self-evident position based on their empirical focus are in fact orientating a positivist stance. Empiricism is after all a theory. In recent years, a variety of theoretical positions have informed historical geographers in different ways, prompting particular interpretive frameworks and concerns when in the archives. Marxism, for example, has facilitated the mining of archives with a view to revealing such issues as historical class power relations, and the structural framework of mobility in capitalist economies. Feminism has effected a concentration on the gendered, racialized and sexualized spaces of the past, which in turn has raised important questions of historical geographical knowledge production, epistemology and research methods. Postcolonialism, too, has been instrumental in problematizing totalizing discourses that are dominant in official government archives. Postcolonial interpretations of such sources have revealed the fallacy of scripted senses of natural or scientific order to societies; an order in fact dependent on power, politics and prejudice.

Narration

The endgame of historical research is to (re)present a knowledge, to resurrect a world from the past, to write its story. Once we have done our research, rendering visible the historical geographies at the heart of our inquiry can be a wonderfully fulfilling exercise, full of considered thought, passion and empathy. *Narration*, however, is not unproblematic, procrastination aside.

To begin with, it is important to remember that our 'representations' are not 'regularities of nature' but rather 'conventions of a situated-geographic-imagination' (Duncan and Ley, 1993: 13). In other words, our story is a *partial truth*, reflective of our knowledge community, theoretical position, methodological techniques and idiosyncratic choices made in narration. As David Wishart (1997: 114, 116) reminds us, 'there is no objective way of judging the "truth value" of a narrative', with the 'facts' being determined 'as much by the narrative as the other way around'. Moreover, if we accept the unavoidable selectivity and subjectivity of the narrative process, 'we might be able to put the objectivist fallacy to rest, opening the possibility for many legitimate interpretations of the past' (Wishart, 1997: 117).

Since the 1970s, geography has witnessed a strong critique of representation, heralding the so-called *crisis of representation* in an ostensibly more relative and fluid postmodern world. More recently, geography has seen a rise of *non-representational* research strategies that seek to negate the primacy of *representations* in understanding space, and focus instead on the everyday *practices* and *performativity* of social productions (Thrift, 1996; Somdahl-Sands, 2008). As Ola Söderström (2005: 14) asks, '[d]oes this mean that the era of representation in geography has come to an end'? 'Certainly not', according to Söderström. For him, the critique of representation may not have impacted on some geographers, who continue to model their writing on a 'Cartesian mirror conception of knowledge' in 'producing "truthful" and univocal images of the (geographical world)'; however, on the whole, it has resulted in geographers increasingly 'analysing the interplay between different forms of representations of space – in maps, photographs, cinema, etc. – and fields of practice, such as patterns of behaviour' (Söderström, 2005: 14). The important prompt here for historical geography is to not use representational sources unproblematically or privilege their explanatory role in narration. Evidence of everyday practices, performances and spectacle may be difficult to find in the historical record, but they can add richly our understanding of the functioning of historical spaces.

In writing the worlds of the past, a key strength that historical geography has orientated over time is its ability to situate *localized* research in *broader* contexts (a key concern for anyone embarking on a research project): 'Historical geography has a long tradition of locating local studies within broader processes operating at wider spatial scales, of paying attention to both the specificity of the local and the wider economic, cultural and political processes' (Nash and Graham, 2000: 1). Situating the localized contexts of our representations does not lead to mere relativism or to a defeatist conception of historical inquiry. Andrew Sayer and Michael Storper (1997: 11) have argued, for instance, that a call for sensitivity to context is not 'by any means equivalent to a plea for particularism or relativism', and that 'it is possible to conceive of situated universalism as a form of normative theory'. Gary Bridge (1997: 633, 638), too, has shown that 'local or situated knowledge and culture can be universally understood, but in a way that is implicated in the constitution of situated knowledges'; he suggests that 'it is time we released the conceptions of conscious social action that enable us to realise the partiality and localness of our knowledge and make them explicit' (see also Katz, 1996).

Conclusion

In thinking through the narrative practice of historical representation, the chapter concludes by considering the degree to which we can hope to make known our *positionality* to the reader of our research. Ian Cook (2005: 22) argues that 'researchers' identities and practices make a big difference', and given the 'politics and ethics' of research and representation, he poses the question 'why not be more reflexive?' in making known our biases and impartiality. David Wishart (1997: 115) wonders if this is 'achievable in any meaningful way'. Certainly, the recognition of the selectivity and subjectivity of historical representation has led scholars to situate their research position and context. Like Cook, Mona Domosh and the late Denis Cosgrove (1993: 37) impel us to 'explicitly recognise our personal and cultural agendas' at the start of our narratives. Stating a bias, however, as Wishart reminds us, is 'a lot more complicated than is often suggested', given that a 'full exegesis would require an autobiography, and that's hardly an unbiased genre' (1997: 115). Wishart asserts that we should credit the reader with being capable of judging the analysis on its own merits, without excessive prompting from the author. The reader, he says, 'will do so in any case' (1997: 115). This is not, of course, to say that we should not contextualize our situated historical representations (as argued throughout this chapter). Situating our knowledge productions, outlining the methodological concerns and signalling the broader theoretical currents is ultimately key to narrating critical, engaging and relevant historical geographies.

KEY POINTS

- Recent years have witnessed new perspectives and themes within historical geography that have been consonant with a broader concern in human geography for theorizing the production of knowledge about nature and culture.
- The notion of *situated knowledges* of the geographical field has resulted in work in historical geography becoming more differentiated, contextualized and self-reflexive of the theoretical and methodological production of knowledge.
- The development of a wider thematic field has coincided with the emergence of a broader delineation of what constitutes valid sources; whose diversity has in turn prompted the employment of new techniques for their effective examination.
- For all sources, thinking through questions of accuracy, authenticity, political and ideological underpinnings and power relations is an essential part of doing historical geography; broader theoretical concerns also form an integral part of interpreting sources.
- In writing, our story is always a *partial truth*, reflective of our knowledge community, theoretical position, methodological techniques and idiosyncratic choices made in narration.

- Situating knowledge productions, outlining methodological concerns and signalling broader theoretical currents are key requirements in narrating critical, engaging and relevant historical geographies.

References

Aitken, S. (2005) 'Textual analysis: reading culture and context', in R. Flowerdew and D. Martin (eds) *Methods in Human Geography* (2nd edn). Harlow: Pearson, pp. 233–249.

Baker, A.R.H. (1997) '"The dead don't answer questionnaires": researching and writing historical geography', *Journal of Geography in Higher Education*, 21(2): 231–243.

Black, I.S. (2003) 'Analysing historical and archive sources', in N.J. Clifford and G. Valentine (eds) *Key Methods in Geography*. London: Sage, pp. 475–500.

Bridge, G. (1997) 'Guest editorial essay: towards a situated universalism: on strategic rationality and "local theory"', *Environment and Planning D: Society and Space*, 15(6): 633–639.

Cloke, P., Cook, I., Crang, P., Goodwin, M., Painter, J. and Philo, C. (eds) (2004) *Practising Human Geography*. London: Sage.

Cook, I. (2005) 'Positionality/situated knowledge', in D. Atkinson, P. Jackson, D. Sibley and N. Washbourne (eds) *Cultural Geography: A Critical Dictionary of Concepts*. London: I.B.Tauris, pp. 16–26.

Cosgrove, D. and Domosh, M. (1993) 'Author and authority: writing and the new cultural geography', in J. Duncan and D. Ley (eds) *Place/Culture/Representation*. London: Routledge, pp. 25–38.

Duncan, J. (1990) *The City as Text: The Politics of Landscape Interpretation*. Cambridge: Cambridge University Press.

Duncan, J. (1999) 'Complicity and resistance in the colonial archive: some issues of method and theory in historical geography', *Historical Geography*, 27: 119–128.

Duncan, J. and Ley, D. (1993) 'Representing the place of culture', in J. Duncan and D. Ley (eds) *Place/Culture/Representation*. London: Routledge, pp. 1–21.

Hannam, K. (2002a) 'Using archives', in P. Shurmer-Smith (ed.) *Doing Cultural Geography*. London: Sage, pp. 113–120.

Hannam, K. (2002b) 'Coping with archival and textual data', in P. Shurmer-Smith (ed.) *Doing Cultural Geography*. London: Sage, pp. 189–197.

Harley, J.B. (1988) 'Maps, knowledge and power', in D. Cosgrove and S. Daniels (eds) *The Iconography of Landscape: Essays on the Symbolic Representation, Design and Use of Past Environments*. Cambridge: Cambridge University Press, pp. 277–312.

Harley, J.B. (1989) 'Historical geography and the cartographic illusion', *Journal of Historical Geography*, 15(1): 80–91.

Henderson, G. (1998) 'Review article: "Landscape is dead, long live landscape": a handbook for sceptics', *Journal of Historical Geography*, 24(1): 94–100.

Katz, C. (1996) 'Towards minor theory', *Environment and Planning D: Society and Space*, 14: 487–499.

Keighren, I.M. (2012) 'Fieldwork in the archive', in R. Phillips and J. Johns (eds) *Fieldwork for Human Geography*. London: Sage, pp. 138–140.
Lorimer, H. (2003) 'Telling small stories: spaces of knowledge and the practice of geography', *Transactions of the Institute of British Geographers* (new series), 28(2): 197–217.
Lorimer, H. and Spedding, N. (2005) 'Locating field science: a geographical family expedition to Glen Roy, Scotland', *British Journal for the History of Science*, 38(1): 13–33.
Morrissey, J. (2004) 'Contours of colonialism: Gaelic Ireland and the early colonial subject', *Irish Geography*, 37(1): 88–102.
Nash, C. and Graham, B. (2000) 'The making of modern historical geographies', in B. Graham and C. Nash (eds) *Modern Historical Geographies*. Harlow: Prentice Hall, pp. 1–9.
Ogborn, M. (1996) 'History, memory and the politics of landscape and space: work in historical geography from autumn '94 to autumn '95', *Progress in Human Geography*, 2(2): 222–229.
Ogborn, M. (2003) 'Finding historical data', in N.J. Clifford and G. Valentine (eds) *Key Methods in Geography*. London: Sage, pp. 101–115.
Sayer, A. and Storper, M. (1997) 'Guest editorial essay. Ethics unbound: for a normative turn in social theory', *Environment and Planning D: Society and Space*, 15(1): 1–17.
Shurmer-Smith, P. (2002) 'Reading texts', in P. Shurmer-Smith (ed.) *Doing Cultural Geography*. London: Sage, pp. 123–136.
Söderström, O. (2005) 'Representation', in D. Atkinson, P. Jackson, D. Sibley and N. Washbourne (eds) *Cultural Geography: A Critical Dictionary of Concepts*. London: I.B.Tauris, pp. 11–15.

Somdahl-Sands, K. (2008) 'Citizenship, civic memory, and urban performance: *Mission Wall Dances*', *Space and Polity*, 12(3): 329–353.
Thrift, N. (1996) *Spatial Formations*. London: Sage.
Wishart, D. (1997) 'The selectivity of historical representation', *Journal of Historical Geography*, 23(2): 111–118.
Withers, C.W.J. (2002) 'Constructing "the geographical archive"', *Area*, 34(3): 303–311.

FURTHER READING

Duncan, J. (1999) 'Complicity and resistance in the colonial archive: some issues of method and theory in historical geography', *Historical Geography*, 27: 119–128.
Nash, C. and Graham, B. (2000) 'The making of modern historical geographies', in B. Graham and C. Nash (eds) *Modern Historical Geographies*. Harlow: Prentice Hall, pp. 1–9.
Ogborn, M. (2003) 'Finding historical data', in N.J. Clifford and G. Valentine (eds) *Key Methods in Geography*. London: Sage, pp. 101–115.
Wishart, D. (1997) 'The selectivity of historical representation', *Journal of Historical Geography*, 23(2): 111–118.

INDEX

Page numbers in *italics* refer to figures.

Index